Mechatronics for the Evil Genius

Evil Genius Series

123 Robotics Experiments for the Evil Genius

Electronic Gadgets for the Evil Genius: 28 Build-It-Yourself Projects

Electronic Circuits for the Evil Genius: 57 Lessons with Projects

123 PIC® Microcontroller Experiments for the Evil Genius

Mechatronics for the Evil Genius: 25 Build-It-Yourself Projects

50 Awesome Automotive Projects for the Evil Genius

Solar Energy Projects for the Evil Genius: 16 Build-It-Yourself Thermoelectric and Mechanical Projects

Bionics for the Evil Genius: 25 Build-It-Yourself Projects

MORE Electronic Gadgets for the Evil Genius: 28 MORE Build-It-Yourself Projects

Mechatronics for the Evil Genius

NEWTON C. BRAGA

McGraw-Hill
New York   Chicago   San Francisco   Lisbon
London   Madrid   Mexico City   Milan   New Delhi
San Juan   Seoul   Singapore   Sydney   Toronto

The McGraw·Hill Companies

Library of Congress Cataloging-in-Publication Data

Braga, Newton C.
 Mechatronics for the evil genius / Newton C. Braga
 p. cm.
 ISBN 0-07-145759-3
 I. Mechatronics. I. Title.

TJ163.12.B72 2005

2005051094

Copyright © 2006 by The McGraw-Hill Commpanies, Inc. All rights reserved. Printed in the United States of America. Except as permitted under the United States Copyright Act of 1976, no part of this publication may be reproduced or distributed in any form or by any means, or stored in a data base or retrieval system, without the prior written permission of the publisher.

7 8 9 0   QDB/QDB   15 14 13

ISBN 0-07-145759-3

The sponsoring editor for this book was Judy Bass and the production supervisor was Pamela A. Pelton.
It was set in Times Ten by MacAllister Publishing Services, LLC.
The art director for the cover was Anthony Landi.

This book is printed on acid-free paper.

McGraw-Hill books are available at special quantity discounts to use as premiums and sales promotions, or for use in corporate training programs. For more information, please write to the Director of Special Sales, McGraw-Hill Professional, Two Penn Plaza, New York, NY 10121-2298. Or contact your local bookstore.

Information contained in this work has been obtained by The McGraw-Hill Companies, Inc. ("McGraw-Hill") from sources believed to be reliable. However, neither McGraw-Hill nor its authors guarantee the accuracy or completeness of any information published herein, and neither McGraw-Hill nor its authors shall be responsible for any errors, omissions, or damages arising out of use of this information. This work is published with the understanding that McGraw-Hill and its authors are supplying information but are not attempting to render professional services. If such services are required, the assistance of an appropriate professional should be sought.

Contents

About the Author	vi
Preface	vii
Acknowledgments	ix
Section One — Preparing the Reader	1
Section Two — The Technology Used in the Projects	5
Section Three — The Projects	11
PROJECT 1 — Mechatronic Race Car	11
PROJECT 2 — RobCom: A Combat Robot	22
PROJECT 3 — Using a PWM Motor Control	33
PROJECT 4 — Ionic Motor	40
PROJECT 5 — Experimental Galvanometer	53
PROJECT 6 — Experimenting with Electromagnets	61
PROJECT 7 — Electronic Potentiometer	67
PROJECT 8 — Experiments with Eolic Generators	73
PROJECT 9 — Electronic Cannon	83
PROJECT 10 — Experiments with Lissajous Figures Generated by a Laser	93
PROJECT 11 — Analog Computer	106
PROJECT 12 — Touch-Controlled Motor	118
PROJECT 13 — Mechatronic Elevator	127
PROJECT 14 — Stepper Motor Control	136
PROJECT 15 — Magic Motion Machine	146
PROJECT 16 — Test Your Nerves	152
PROJECT 17 — Robot with Sensors	158
PROJECT 18 — SMA Experimental Robotic Arm	164
PROJECT 19 — Position Sensor	172
PROJECT 20 — Light-Beam Remote Control	177
PROJECT 21 — Mechatronic Airboat	183
PROJECT 22 — Coin Tosser	187
PROJECT 23 — The Challenge of the Mechatronic Timer	191
PROJECT 24 — Experimental PLC	195
PROJECT 25 — The Mechatronic Talking Head	201

About the Author

Mr. Braga was born in Sao Paulo, Brazil, in 1946. His activities in electronics began when he was only 13 years old, at which time he began to write articles for Brazilian magazines. At age 18, he had his own column in the Brazilian edition of *Popular Electronics*, where he introduced the concept of "electronics for youngsters."

In 1976, he became the technical director of the most important electronics magazine in South America, *Revista Saber Eletronica* (published at that time in Brazil, Argentina, Colombia, and Mexico). He also became technical director of other magazines of *Editora Saber,* such as *Eletronica Total,* and became the technical consultant for the magazines *Mecatronica Facil, Mecatronica Atual*, and *PC & CIA*.

Mr. Braga has published more than 90 books about electronics, mechatronics, computers, and electricity, as well as thousands of articles and electronics/mechatronics projects in magazines all over the world (U.S., France, Spain, Japan, Portugal, Mexico, and Argentina among others). Many of his books have been recommended at schools and universities around the world and been translated into other languages, with sales of more than 3 million copies worldwide.

The author currently teaches mechatronics at Colegio Mater Amabilis, is a consultant for distance learning, and is engaged in educational projects in his home country of Brazil. These projects are directed at the introduction of electronics and mechatronics in middle schools as well as the professional training of workers and teachers who need enhanced knowledge in the fields of electronics, mechatronics, and technology. The author now lives in Guarulhos (near Sao Paulo, Brazil) with his wife and 15-year-old son.

Preface

This book doesn't pretend to be a complete resource for the mechatronics evil genius, but it certainly offers a large assortment of useful information and ideas for projects not found elsewhere.

The purpose of this book is not only to teach many tricks and techniques used to build mechatronic and electronic devices, but also to provide ideas and complete projects that can be easily duplicated using low-cost and easy-to-find parts.

The audience for *Mechatronics for the Evil Genius* includes beginner, intermediate, and advanced builders who want new ideas for projects, as well as educators who want to introduce the use of technology to their schools. Of course, the most important reader is the evil genius who can make incredible things using his or her imagination, skills, and some parts gathered from old equipment and appliances, stolen from a younger sibling's toys, or bought from a local electronics parts dealer.

If you think that it's impossible to build interesting things using simple materials and technology, you are wrong. Three types of technology are used to build electronic and mechatronic projects:

- The simplest, or "traditional," technology incorporates electrical parts for motors, cells, and passive components. This technology can be understood even by children in elementary school. You can implement some interesting projects with this technology by simply using your imagination, and we will present some projects using these simple devices in this book.
- Intermediate technology is the kind that uses something more advanced than the passive components of the traditional physicist, such as semiconductors (diodes, transistors, *silicon controlled rectifiers* [SCRs], *light-emitting diodes* [LEDs]) and some *integrated circuits* [ICs], but not so advanced as microprocessors, *very large scale integration* (VLSI) chips, *digital signal processors* (DSPs), and much more.

    The great advantage of using intermediate technology is that it is accessible to all. Discrete components such as transistors, resistors, and diodes can be easily handled, which is important in order to reveal hidden vocations and talents. You will not need special tools to handle such components, as they are vigorous enough to resist the evil genius who is not experienced with the use of tools.
- The advanced technology is the type you can find in all modern appliances such as cellular phones, DVDs, computers, pagers, video games, and *global positioning systems* (GPSs). Although these appliances use very complex chips, they are all based on the same principles of operation. What differs is the number of functions and components.

You can make a simple radio using three or four components, but a high-tech radio requires a microprocessor with a million components. The most important goal for the evil genius is that he or she can build different things, appliances not commonly found but seen only in movies, on TV, or in science fiction magazines. Using low-cost parts and easy technology, the reader can build simple robots, shock machines, race cars, remote controls, and much more with the same approach shown in TV programs such as those on the Discovery Channel.

You can do all that, and we intend to give you some of the necessary tools, ideas, and techniques in the following chapters. You only have to complete these items with your imagination—the super imagination found only in a true evil genius. This book is divided into three parts:

- In the first part, we will prepare the reader to be a real mechatronics evil genius. We will explain what to do and how to handle electronic and mechanical devices used in these projects. We will also dedicate space to educators who want to reveal an evil genius or two among their pupils by building projects and making experiments with them. The educators will see how easy it is to link many of our projects with their science classes, using the projects as cross themes for the science curriculum.

    For readers who want to take a more serious approach in the hope of becoming an amateur scientist, we will discuss the scientific method. This will help the reader to go further with the projects, developing new and more advanced things from the ideas provided in this book. It is an approach that will involve the imagination of the advanced evil genius.

- In the second part of the book, we will show the reader how to build electronic circuits, work with electronic components and devices, mount materials, use tools, and solder. Because the evil genius tends to be daring, sometimes going too far with the projects and their use, this book details how to avoid dangerous situations when working with technology and how to work and play safely.
- In Part 3, we find the projects, 25 of them, chosen from the author's large collection of experiments, many specially created for the reader of this book, the real evil genius.

The projects are all complete; they have all the information needed to build a basic version that works. A brief description of the project explains how it is supposed to look and work upon completion. Following this section, the reader will find the operation principles and details of how to mount the materials. A complete parts list makes it easy for the reader to gather all the parts needed for building. After that, we include directions for making any adjustments, adding additional circuits, upgrading the project, building variations, or conceiving new projects based on the same principles.

Rounding out each project are ideas for educators, linking projects to the subjects taught in school, as well as additional information. This approach makes the book easy to use as a large reference for mechatronics projects.

We hope that as a reader and potential evil genius your face will light up and your eyes will be full of mischief from the ideas provided in this book. Enjoy!

Newton C. Braga

Acknowledgments

I would like to thank all the people who helped make this book possible:

Jeff Eckert—my book agent who helped me with all the bureaucratic procedures involving the production of the book.

Carlos Eduardo Portela Godoy and Marcelo Portela Godoy—who gave me support when working with my mechatronics pupils at Colegio Mater Amabilis in Guarulhos (Brazil), revealing among them many an evil genius.

Helio Fittipaldi—who allowed me to use many illustrations and photos from articles I had published in his magazines *Mecatronica Facil* and *Eletronica Total*.

Edson de Santis—my great friend who supplied me with many parts of the components I have used in the projects described in this book.

Alexandre Costa Berbel—who helped me monitor how many of the projects described in this book were used by a large number of schools in the city where I live and many others where he supports the use of technology in education.

My wife Neuza and my son Marcelo—who have both been supportive of my efforts.

Newton C. Braga

Section One

Preparing the Reader

What Is Mechatronics?

So you want to be an evil genius in mechatronics? Before you start with the fantastic projects described in this book, can you answer a simple question?

Do you really know what mechatronics is?

If you are not sure about the correct answer to this question, it would be better to first learn something about this new fantastic science.

When you saw the word "mechatronics" on the cover of this book, the first thing that probably came to mind was building robots. Robots are the main products of mechatronics, and robots are very popular these days. Just look at their presence in movies, toys, and books. Character names like R2-D2, Asimo, Aibo, and movie names like Asimov's *I, Robot* are known by everyone.

These increasingly popular mechanical creatures are part of a branch of science known as *robotics*, and robotics is *part* of mechatronics. So, to know more about mechatronics we must return to the origin of robotics and even the automatic machine using no more than common mechanics.

History

The idea that machines can perform difficult and repetitive tasks, acting as servants (and freeing humans from the tasks), originated in ancient Greece. Traces of movable statues were found in the first century B.C. One individual in particular to be acknowledged is Hero, who was from Alexandria and conducted experiments with mechanical birds. Accounts of Cresibus, a Greek engineer, claim he built organs and water clocks with movable figures.

In 730 A.D., the Swiss clock maker, Pierre Jacquet-Droz, built three mechanical tools that could play music on an organ, draw simple figures, and write. Later, but not so long ago, Tesla built a remote-controlled submarine.

But it was the Czech novelist Karel Kapec, in his book *R.U.R., Rassum's Universal Robots*, who first used the term *robot*. In the book, he describes mechanical servants doing all the things a man could do. The word robot simply means *worker* in Czech.

Looking further ahead in time, the idea of mechanisms, though not necessarily humanlike mechanisms, performing tasks for humans has not disappeared. With the development of new technologies such as electronics, cybernetics, and artificial intelligence, a new science named mechatronics appeared.

Mechatronics can be defined as "the synergistic integration of mechanics, electronics, and computer technology." It can be classified as a subject of cybernetics. Figure 1.1 shows a graphic where we place mechatronics as an independent science.

Mechatronics and robotics have many points in common. Both depend on electronic and mechanical parts, and their projects operate following the same principles.

Looking at the curriculum of many mechatronics courses (also called industrial automation), we can

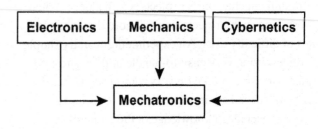

Figure 1.1 *How mechatronics interacts with other sciences.*

see that the main subject is robotic technology, focusing on industrial robots or automatic manufacturing machines.

This book doesn't intend to be a course in mechatronics, but it is a practical approach to the subject. What we intend to do with this book is show the reader how to have fun and learn about such traditional sciences as physics, biology, and others by using mechanics and electronics or, if you prefer, mechatronics.

Mechatronic Tools and Principles

At this point, we believe that if the reader wants to build the projects described in this book, it is necessary to be prepared to handle tools and know the basic principles behind the two sciences comprising mechatronics. It is also clear that by visiting many model, toy, and virtual stores the reader can find complete mechatronic projects ready to be used. These store-bought projects may be even more complicated or less expensive than the projects described in this book.

However, the real evil genius makes discoveries through his or her own efforts, inventing new devices and learning the theory necessary to build them. The real evil genius doesn't buy an operating robot or a kit; he builds his own robot. The real evil genius doesn't look for an ionic motor in science stores; she creates her own version. That is the great difference in what makes this book ideal for a true evil genius.

Barriers to Be Crossed

Building mechatronics projects is a challenge. When observing such projects, one will note that many levels of difficulty exist. An experimenter can try simple projects using very easy-to-handle and easy-to-find parts, or one can take on very complicated ones that require special components, special tools, and a highly specialized knowledge of the technology involved.

Many readers are familiar with the use of delicate components such as the ones found in electronic circuits or small mechanical devices, whereas other readers are not. For those who are not familiar with the technology used in the projects, some barriers must be crossed. Starting with simple projects, the inexperienced reader can acquire the necessary skills to graduate to more complex projects in relatively little time. This means that we will present the reader with both easy and intermediate projects, from the simplest for anyone with a minimum knowledge of technology and tools to others that require much more experience to build. Information about each project's degree of difficulty will be given.

The reader must also have the necessary tools to work with these delicate parts. When working with electronic and mechanical parts, the reader must have specialized tools as well as sufficient knowledge of how to use them. In the next few pages, we will give some important information about these tools.

To complete this section, the reader must answer a final question: Can I use the projects for any purpose other than simply having fun? The projects described in this book are not simply toys. They can be used for other ends. The amateur scientist can use the ideas given by these projects to discover new things. Practical circuits and configurations can be implemented to build simple research tools used in experiments involving chemistry, physics, and even human sciences.

Educational Electronics, Robotics, and Mechatronics

One of the most interesting applications of the projects in this book is developing an evil genius. The projects can easily be included as technological cross-theme material complementing school curriculum and adding to students' routine activities. Remember that elementary schools are currently trying to include technology in their curriculum.

Although many believe that technology means "using a computer," today's technology is present in all parts of our world. Technology starts with the simple lamp you turn on and off, and it includes such electronic and mechatronic domestic appliances as VCRs, TVs, telephones, and high-tech devices such as robots, radars, cellular phones, and wireless computer links. Technology is even more prevalent in some locations. Just visit an airport, a shopping center, or a bank.

The basic idea of educational electronics, robotics, and mechatronics has been adopted here in my home country of Brazil (and in many other countries). Using technology with many of the projects described in this book, I not only entertain readers, but provide the educator with the opportunity to use the projects in order to link the basic sciences (physics, chemistry, and biology) with higher-end technology.

Throughout this book, I show photos of my mechatronics pupils using the projects, such as those that mount a combat robot or mechatronic race car. Students not only have to do research in physics (mechanics) to reduce friction, create the fastest cars, and explain how they work, but they are also challenged to put all their evil genius skills to work.

The projects in this book are created using a central part that can stand alone as a complete device, and parts may be added to create new projects. Experienced and innovative teachers can use additional circuits and creativity to produce new projects and to complement or supplement a curriculum.

Cross Themes

Teachers and educators are very familiar with cross themes. When working with the items of a common science curriculum, it is a good idea to insert activities that are not directly indicated in the curriculum but can be used to reinforce the topic being studied.

Cross themes are very important in today's education and are highly recommended by educational authorities. They make it easier for students to understand many scientific principles that are not very clear when simply using a blackboard or other traditional teaching tools.

The Scientific Method

Technology is the practical application of science. In earlier times, technology grew out of direct experience with the properties of things and the techniques for manipulating them. Today technology also depends on a vast array of formulas and theories.

Engineering (like mechatronics) is the systematic application of scientific knowledge in developing technology. Engineering has grown from a craft to a science in itself. Therefore, when creating projects in mechatronics, you need to follow the rules of science and technology research. These rules are described in the scientific method.

The scientific method is the process by which scientists construct a reliable, consistent, and nonarbitrary representation of the world. The scientific method has four steps:

1. Observation and description of a phenomenon
2. Formulation of a hypothesis to explain the phenomenon
3. Use of the hypothesis to predict the existence of other phenomena
4. Testing the predictions through independent means

To do an experimental science project, these steps must be followed:

1. Make the initial observation.
2. Gather information.
3. Name and rationalize a purpose for the project.
4. Formulate a hypothesis.
5. Design an experimental procedure to test your hypothesis.
6. Gather material and equipment.
7. Do the experiment and record the data.
8. Make calculations and semmarize results.
9. Try to answer the original questions.

It is important to observe that the scientific method has different forms. Physical scientists do experiments to gather numerical data from which relationships are derived. From the results, conclusions are made. Descriptive scientists, such as anthropologists, use information gathered by observation and interviewing.

In mechatronics, the reader must follow the procedures used for all experimental science projects. One must collect data, form a hypothesis, gather material, and create a project that can be used to confirm one's hypothesis and/or that can be used in some practical application.

Choosing a Project

The projects found in this book have different levels of difficulty. It is up to the reader to analyze all the projects and decide if the original version or an upgraded version can be done. Take into consideration the following when choosing a project:

- Is the project at an appropriate level of difficulty in respect to your knowledge of mechatronics?
- Are the parts clear enough to prevent difficulty when mounting?
- Do you have the necessary tools to handle all parts of the project?
- Do you know how to use the tools needed to build the project?
- Are you sure that you can find all the parts at your supplier?
- If you need the help of a more experienced person, is this person readily available?

Take into account these points so you don't have any unpleasant surprises when mounting a particular project.

Section Two

The Technology Used in the Projects

Most of the projects described in this book can be used individually. Wherever possible, the circuits have been designed so that they can be grouped with one or more other projects. By replacing some components, they can be altered to execute new functions with other aims.

How to Mount

Consumer appliances use machines to assemble small parts, starting with a technology called *surface mount technology* (SMT), which employs very small parts. Without the use of special tools, we could not handle these parts, and it is very difficult to mount any project using them. Today's technology is different from the basic technology that used old parts as tubes and only electric components as lamp switches, fuses, and other parts.

For the typical reader, the ideal is to begin by using intermediate technology. The components are a bit larger and can be handled with common tools. You might think these devices would be slower, but this is actually not the case. Larger components do the same tasks done by the small *surface mount devices* (SMD). They simply need more space in a board to be placed. They are ideal to work with, particularly if you are not familiar with the tools and do not yet have the skills to handle very small parts. These are the size of the components we intend to use in the projects described in this book.

So, the first thing you have to do before choosing a project to build alone, with your friends, with pupils, or for your work in a science fair is to learn something about the tools you will use and the components you will assemble in your project. The first primary skill you need to know is how to solder.

The Tools

Before starting any project, particularly if it is your first, you need to know which tools to use when working with electronic circuits.

How to Mount Components

Electronic components are comprised of small pieces that need some support to be kept in place and wired in a circuit. Several techniques can be used to mount components. The simplest way to place a component in a circuit is to use a terminal strip, as shown in Figure 2.1.

The components are soldered to the terminals, and interconnecting wires are soldered in the

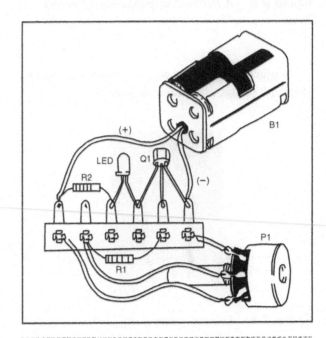

Figure 2.1 *The terminal strip can be used as a chassis for simple projects.*

5

correspondent terminals. The way the components are placed and wired determines what the circuit will do.

This is not the best way to mount a project, but it has the advantage of being very simple. This method does not require special tools or resources. We will use this technique to assemble the projects specified for beginners or for students who are not familiar with more advanced techniques, such as the ones that use *printed circuit boards* (PCBs).

Terminal strips can be easily found and improvised by using a plastic or wooden strip where small terminals, nails, or other metallic pieces can be attached as shown in Figure 2.2.

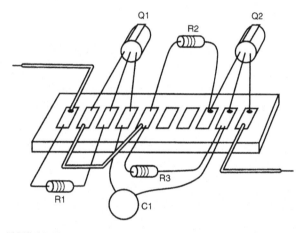

Figure 2.2 *A terminal strip made in a piece of PCB.*

Another way to wire a circuit is to use a terminal strip with screws, as shown in Figure 2.3.

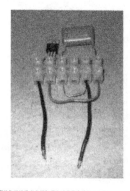

Figure 2.3 *Using a terminal strip with screws to assemble a simple circuit.*

The advantage of using these terminal strips is that the components don't need to be soldered. The disadvantage is that you must take special care with the connections to place them firmly. Any bad contact can affect the circuit performance or even prevent it from functioning. Finally, many experimental projects can be mounted on solderless boards, such as the one shown in Figure 2.4.

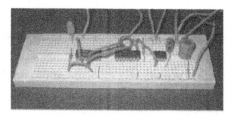

Figure 2.4 *Solderless board.*

The components' terminals are placed into holes where metallic terminal strips interconnect them according to a predetermined pattern. The advantages of this technique are that you don't need to solder the components and they can be reused in other projects. It is also very easy to replace components, making it possible to achieve the best values for the desired performance.

In our projects we will use all levels of difficulty for the techniques described previously. The level chosen depends on the project and will be suggested according to the degree of difficulty for each project.

Before showing readers how to solder, let's take a look at the PCB and see how it is used.

The Printed Circuit Board (PCB)

The small components used in electronic equipment can't stand alone without physical support. They need some kind of structure to keep them fixed and, at the same time, allow them to provide the electric connection with the rest of the circuit.

Opening any electronic equipment, the reader will find that the small parts or electronic components are mounted on a special board of fiber or other insulating material. This support or chassis for the components is a PCB.

Figure 2.5 *Common PCBs.*

The board, as shown in Figure 2.5, is made of an insulating material, and copper strips are printed on one or both sides.

The copper strips act as wires conducting the currents from one component to another. The pattern of the strips is determined by the function of the circuit. All the strips are planned before the manufacturing process to provide the necessary connections between components and the desired function. This means that a PCB produced to receive, or hold, components that form a radio can't be used to mount the components required in a TV circuit or other equipment.

As seen in Figure 2.6, the small components are soldered onto the board so that their terminals make contact with the copper strips.

In some cases, when the components are very small and programmed to be fixed by automatic machines as in the SMDs, they can be placed on the board on the same side as the strips (see Figure 2.7).

As explained earlier, the SMD is not recommended for readers who are looking for simple and accessible projects.

Installing components on a PCB is a delicate operation that any reader who wants to work with electronic mounting circuits must know. Because the components are small and extremely fragile, special techniques are required for this work.

The solder used in an electronic assembly is an alloy formed by 60 percent tin and 40 percent lead with some rosin. This kind of solder is typically referred to as a transistor solder, radio-TV solder, or 60-40 solder.

When heated to about 273 degrees Celsius, the solder melts to the circuit board's terminal components, thereby fixing it to the board. This melting simultane-

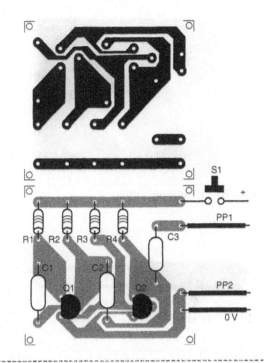

Figure 2.6 *Islands of solder connect the components' terminals to the PCB.*

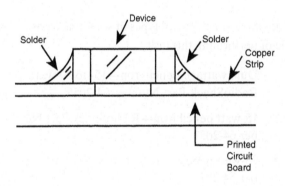

Figure 2.7 *SMD components are soldered directly to the copper lines.*

ously provides electric contact with the copper strips or similar components. When working with mounted or replacement components, the reader will need solders and a soldering iron. Solders can be bought in small quantities, as shown in Figure 2.8. A soldering iron is shown in Figure 2.9.

A 25- to 40-watt soldering iron with a shiny tip is recommended when working with the small electronic components found in circuits. Of course, the builder can use a heavy-duty soldering iron to remove or place larger components such as those found in some electric and electronic projects.

Figure 2.8 *Common solder.*

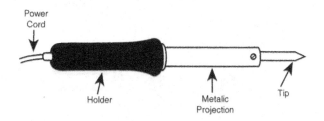

Figure 2.9 *Common soldering iron, intended for the builder who wants to build the projects described in this book.*

RadioShack has several types of soldering irons capable of performing this task:

- Dual wattage pencil iron—15- to 30-watt RadioShack Part Number 64-2055

- 15-watt pencil iron—RadioShack Part Number 64-2051

- 25-watt pencil iron—RadioShack Part Number 64-2070

Soldering is a simple operation, and all technicians should be familiar with this process. However, those just beginning their study of electronic technology may not be.

Electronic devices are very fragile and special care must be taken to avoid damaging them. Many electronic devices are easily damaged by excess heat or incorrect soldering techniques. The basic procedure to solder electronic components (remove or install on a PCB) is as follows:

1. Plug in the soldering iron and heat for at least 5 minutes. This ensures that the tip reaches the appropriate temperature for effective soldering.

2. Touch the soldering iron to the work, allowing a very short time for the connection to heat

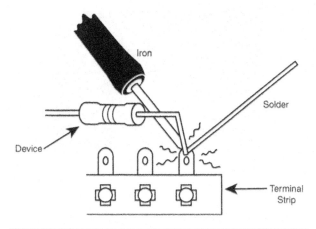

Figure 2.10 *Soldering a component to a terminal strip.*

up. Then touch the solder to the connection, not the iron, as shown in Figure 2.10.

3. You will notice that the solder penetrates every part of the solder joint when melting.

4. Remove the iron and do not move the joint until it has had time to cool. It is relatively easy to discern when the joint has cooled to an appropriate temperature. A peculiar haze will pass over the metal, at which time the technician will know that the joint is cool and strong enough to withstand movement.

Figure 2.11 shows a perfect solder joint and a solder joint with problems. One of the principal causes of problems in electronic equipment is the "cool solder." The solder appears to involve the component, but in actuality no electric contact has been established. Often the joint was not heated long enough to penetrate the metal, an error that allows an isolated layer of moisture or oxide to form between them.

Although the use of the soldering gun (such as the one shown in Figure 2.12) is not forbidden, many

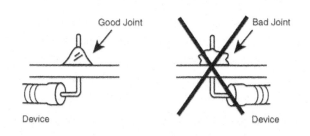

Figure 2.11 *A good soldered joint.*

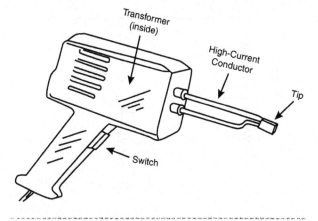

Figure 2.12 *A soldering gun.*

instances exist in electronics in which it is not recommended.

This type of soldering tool has its resistive tip heated by a strong current coming from the low-voltage winding of a transformer. This current can potentially burn more sensitive components, such as *integrated circuits* (ICs) and transistors.

Other Tools

The soldering iron isn't the only special tool needed by readers who intend to work with electronic circuits. Many of the tools used in electric installations or automotive electricity are suitable for working with electronic parts. The reader may in fact have many of these tools readily available at his or her home. On the other hand, many electronic components are small and delicate and thus require special tools and care. The use of improper tools when working with these components can cause serious damage. I suggest having at least some of the following tools:

- Cutting pliers (often called diagonals or dykes), from 4 to 6 inches long
- Chain-nose or needle-nosed pliers with very narrow tips that are 4 to 5 inches long
- Two or more screwdrivers between 2 and 8 inches long
- Crimping tools—a stripper and cutter for 10- to 22-guage wire (RadioShack part number 278-238)
- A precision tool set (10 to 16 tools) with small screwdrivers of hex, common, and Philips types
- Soldering and desoldering accessories such as a desoldering bulb and a soldering iron holder/cleaner
- Extra hands to hold the work as a minivise with vacuum base or a project holder
- Mini-hand drill

Many other tools can be found in electronic and tool catalogs.

Diagrams and Symbols

Schematic diagrams (or simply *schematics*) are used to represent the way many electronic components are interconnected. The builder of electronic projects must be able to not only recognize the components, but also recognize the way they are interconnected by studying a schematic diagram. In a schematic diagram, components are not represented by a real format. They are instead represented by symbols. The reader must therefore become acquainted with a special symbology.

Learning the symbols that are used to represent each component and what they do within a circuit is essential if a reader hopes to master electronics projects. Let's begin by learning how to read a schematic diagram.

Figure 2.13 is a diagram of some simple electronic equipment: the electronic control of a DC motor, used in our first project, the mechatronic race car.

This schematic diagram represents all components by their symbols and in many cases gives the identity, values, and other important information. On the side of each component's symbol is its identification number. This is important because it helps the builder find the component on the PCB or the terminal strip on the inside of the equipment.

For example, all resistors are identified by the letter R followed by the number of the device in the project. This means that many resistors can be identified within a device by such labels as R1, R2, R3, and so on. Capacitors are usually noted by the letter C. The capacitors of a circuit are numbered

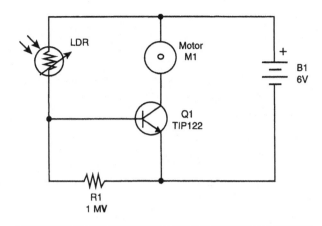

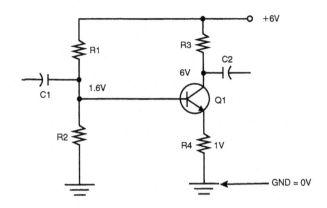

Figure 2.13 *Representing a circuit by symbols.*

Figure 2.14 *The voltages at various points.*

starting from C1, C2, C3, and so on. Transistors can be identified by Q, T, or TR. Thus, the same transistor can be represented as Q1, T1, or TR1.

In many cases, a second number can identify the block or stage in which the component is placed. For example, a resistor in the first stage can be referred to as R101, and a resistor in the second stage as R201. Near the identification symbol, we can also find the value or type of the component.

Resistors have the value of resistance printed on their side (e.g., R1 might have printed on it 1,000 ohms or 1k). If it is a transistor, you might find 2N3906, which means that, when used, the transistor must be replaced with a 2N3906, or when mounted the transistor placed at this location must be a 2N3906. The usual identification number for transistors begins with a 2N, though many manufacturers have begun using alphabetic characters to signify individual brand names, such as TIP for Texas Instruments and MPS or MM for Motorola. Also, a European code uses the letters BC or BD to identify its devices. Additionally, a Japanese configuration uses a combination of numbers and letters such as 2SB, 2SC, or 2SD to indicate transistors.

Depending on the circuit, other important information can be found in the schematic diagram. For example, the voltage at different points and intervals in the circuit can be calculated.

In the example presented, a multimeter measures the voltage at 6 V between A and the ground (normally used as reference or indicated as 0 V) (see Figure 2.14).

Another important piece of information found in the figure is the waveform of a signal passing through a specific point. Procedures for installation, potential problems, and the values for equivalence can also be found within a schematic diagram.

To make it easier for the reader to perform the projects outlined here, we will give figures that detail the construction and list the parts used in addition to providing schematic diagrams. The listing of parts will show that the main characteristic of all components is the color coding used to represent resistors, voltage, current rates, capacitors, transformers, and diodes among others.

Complementary Information

The projects described in this book are simple and don't need special techniques or tools to be assembled. The text provides all the necessary elements to build them. The reader does not need special knowledge of electronics or previous experience with projects.

Although most steps in the projects are not particularly dangerous, those powered from the AC power line contain pointed and cut parts that could potentially cause injury. The reader must therefore take proper precaution when performing these projects in order to avoid injury to themselves and others. Always cover live or dangerous parts when transporting or when the project is not in use.

Section Three

The Projects

The purpose of this section is to present a collection of 25 practical projects for mechatronic applications. To fully understand the projects and allow the reader to customize them to his or her own individual applications, it is important to have read the previous two chapters. If the reader feels he already has a good grounding in this area, it is time to start.

Project 1—Mechatronic Race Car

Purpose

I have adopted this project in my own mechatronics course, designating it as the second practical activity of the year. Students aged 11 to 14 are required to construct a race car in two versions (fan propeller or gears) and enter their finished products in a race.

The car, built with such miscellaneous parts as cards and parts of toys is basically light controlled. A flashlight is used to control the small DC motor that's used as an engine. In the race, the cars are distributed in teams of six to eight and aligned at a starting point. They have to cover a distance from 10 to 20 meters in a straight path and cross the finish line.

Figures 3.1.1 and 3.1.2 show my pupils with their cars at Colegio Mater Amabilis in Guarulhos, São Paulo, Brazil. Since being published in Brazilian magazines, this project has been adopted by schools around the world.

The simple competition challenges students to build the best car (the lightest and fastest). However, the project can also serve an educational purpose. Pupils can learn much about physics, technology, and other sciences while working with others to build a race car.

Figure 3.1.1 *Students in Colegio Mater Amabilis (Guarulhos, Brazil) receiving their prizes after a race (Courtesy Revista Mecatronica Facil).*

Figure 3.1.2 *Example of a race car built by a student of Colegio Mater Amabilis.*

Objectives

There are many objectives when building a race car:

- Learn how to build simple projects by using a terminal strip
- Learn how photocells or photosensors (*light-dependent resistors* [LDRs]) function
- Design a propulsion system that uses gears or a fan
- Learn about friction and how it can be reduced
- Learn how a transistor works as a switch

The Project

As shown in Figures 3.1.3 and 3.1.4, we can build the race car in two different versions: by using a fan as a propeller or by using gears that transfer power from the motor to the wheels. The choice depends on the resources at hand.

In the fan version, the fan can be constructed using a common *compact disc* (CD), cut and bent with the aid of some heat. Although this is the simplest method of construction, the CD is very delicate and many fans break during the competition or even

Figure 3.1.3   *Car using a fan (Courtesy Revista Mecatronica Facil).*

Figure 3.1.4   *Car with gears (Courtesy Revista Mecatronica Facil).*

while being cut and bent to form the fan. It is recommended to have many fans ready to serve as possible replacements during the competition.

Another method is the use of plastic or wooden fans, such as those used in aircraft models. It is up to the reader to find the fan that best suits the project. Certainly, the evil genius will be able to identify the best fan needed to win a competition. Studying characteristics such as the ratio between area and motor speed, diameter, and weight, the evil genius can ultimately identify the fan that will propel his or her car ahead of the competition.

Usually, the gear version is more efficient. However, although the race car may be faster, the issue remains as to how to find the appropriate gears. We suggest using parts of toys or electrical or electronic appliances. These parts should be an accessible gear source for most readers.

The project can be divided into two parts:

- The electronic circuit
- The mechanical part (the vehicle)

As a general rule, inexperienced builders should allot at least 4 hours of work for each part of the project. To make it easier for my students, I prepare kits that include the electronic parts, motors, sensors, and transistors. This is particularly important if you are moderating a competition; it ensures that the competition will not be determined by the materials used, but rather by the skill of the designers.

Rules are provided determining the dimensions of the cars, the power supply, and other items that could potentially make some difference in the race. The evil genius can suggest a competition with his or her neighbors or at his or her school.

How It Works—The Electronic Circuit

The basic electronic circuit is the same for the two versions, so the description of its operation principles is valid for both.

The sensor is an LDR (*cadmium sulfur* [CdS] cell or photoresistor), controlling the base current of a transistor. A Darlington transistor, or high-gain transistor, is used as a switch that controls the current flowing across the motor.

When the LDR receives light, its electric resistance falls, switching the transistor. At this moment, it passes from the off state to the on state, and the current flows to power the small DC motor. See that the transistor acts only as a switch and does not amplify the light. This is important when choosing a light source to be used as a remote control. This means that, when activated by the appropriated light level, the transistor becomes saturated and no additional power can be used to drive the motor. A saturation curve is shown in Figure 3.1.5.

It is important to make clear that it is not the power of the flashlight that excites the LDR and determines the final speed of the car but rather many other mechanical factors. The flashlight is used only to turn on and off the circuit, not to power it. It serves as the remote control, not the power source.

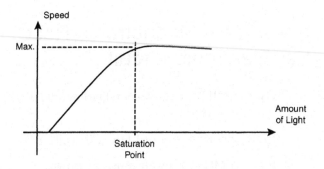

Figure 3.1.5 *After a previously determined point, any increase in the amount of light on the sensor will have no effect on the circuit.*

With the aim of receiving the light only from the flashlight, the LDR is mounted inside a small card tube. Some type of cover should be used to avoid the effects of ambient light prior to the start of the race. One idea is to allow the ambient light to act on the LDR. Therefore, when the race begins, the competitor need only remove the cover from the sensor, allowing the ambient light to switch on the motor.

The power supply for the car is formed by four AA cells, and other cell sizes are not recommended. Large cells, such as C or D cells, do not add power in the same proportion to the rise in the weight they cause, so the speed tends to decrease. Furthermore, the excess of current can overheat the transistor and possibly burn it out. We recommend the use of new alkaline cells on the day of the race. Attaching a small heatsink to the transistor may be necessary in some cases.

How It Works—The Mechanical Part

The chassis is the same for the two versions with small differences only in the propulsion system. Cardboard, plastic, or even light wood can be used to form the chassis, the place where the electronic circuit, motor, and propeller are mounted. The builder is free to make changes in the original project, as shown in previous figures.

Three-wheel versions are accepted. The builder must take care with the alignment of the wheels to be sure that the vehicle will move in a straight line. Remember that the only control you have is in switching the motor on and off.

In the fan-propelled version, the motor is placed on a small platform built with common pieces. An empty inkjet cartridge is used to support the motor in Figure 3.1.1, for instance.

Building the Race Car

Now we can start with the construction of the race car, which is divided in two parts. In the first part, we will describe how to assemble the electronic circuit. In the second part, we will see how to mount the mechanic part of the car.

Electronic Circuit

Figure 3.1.6 shows the complete circuit of the light-operated remote control used in both versions of the race car.

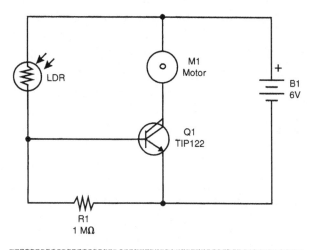

Figure 3.1.6 *Complete schematic diagram of the electronic circuit used to control the DC motor in the race car.*

This circuit can be mounted using a small terminal strip as a chassis, as shown in Figure 3.1.7.

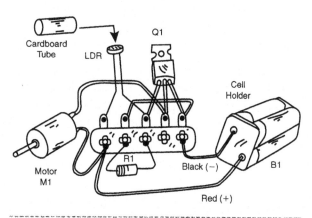

Figure 3.1.7 *A terminal strip can be used to support the small components of the electronic circuit.*

When assembling the circuit you must observe the following:

- The position of the transistor.

- The polarity of the power supply (cell's holder) and the motor. If the motor rotates in a direction that moves the car backward, invert the wires.

- Do not let any of the components' terminals touch each other. This can cause a shortage and risk melting.

Q1 is any Darlington transistor rated to 2 amps or more. If the transistor tends to overheat, attach a small piece of metal to act as a heatsink, as shown in Figure 3.1.8. Remember: The heatsink means more weight added to the car and might reduce its final speed.

Any common LDR can be used in this project. Small round types are ideal because they can be easily placed inside the card tube and improve directional control.

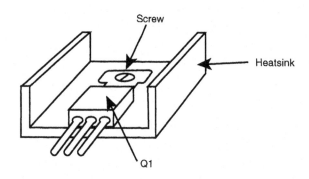

Figure 3.1.8 *Heatsink attached to the power transistor to avoid excess heat.*

Parts List—Mechatronic Race Car Electronic Circuit

Q1   TIP122 or equivalent, Darlington NPN transistor

LDR Photoresistor or LDR (CdS cell)

R1   1 MΩ × 1/8-watt resistor (brown, black, red)

M1   6-volt small DC motor

B1   6 volts of power (4 AA cells and holder)

Cell holder, terminal strip, wires, etc.

Mechatronic Car—Using Fan

Common materials such as cardboard, plastic files, light wood, or metal (aluminum, zinc, etc.) can be used to build the race car's chassis. Figure 3.1.9 shows the basic dimensions of the chassis when using a fan as a propeller.

Figure 3.1.9   *A basic design for the chassis using a piece of cardboard (Courtesy Mecatronica Facil).*

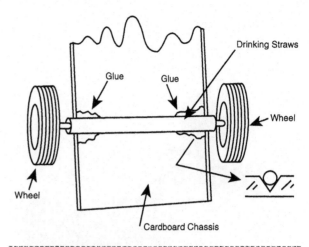

Figure 3.1.10   *Pieces of drinking straws serve as support to the wheels.*

Figure 3.1.11 shows how the fan is built from a CD. Cut it and bend the blades using the heat of a candle or another heat source. Be careful not to burn or break the blades.

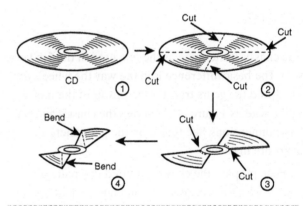

Figure 3.1.11   *CDs can be cut and bent with the aid of some heat from a candle.*

The dimensions given in Figure 3.1.13 represent the average. The builder can change them according to the size of wheels he or she intends to use and according to other design factors. It is important that the organizers of the race limit the dimensions to avoid aberrations in the competition.

Such wheels can be found in toys and can be installed on the race car, as shown in Figure 3.1.10. Small plastic drinking straws are often used as axles to complete this task.

It is important that the straws be long enough to position the wheels firmly, as the car must run in a straight line. The motor with the fan can be installed at the top of an empty cartridge of an inkjet printer. Use glue or another method to fasten it in position.

The fan constructed with a CD is very delicate, breaking easily at any impact. Have several units available to replace them during a race or test. Many fans can break during a competition.

The fan is glued to a small plastic wheel taken from a toy. Be sure that the wheel is centralized. Otherwise, the fan will vibrate when running, either breaking off or taking the car off its straight line path. The batteries and the electronic circuit are placed in the chassis as shown in Figure 3.1.12.

Perform tests to see if the vehicle runs forward when activated. If not, invert the wires to the motor.

Figure 3.1.12 *The race car ready for test (Courtesy Mecatronica Facil).*

Mechatronic Car—Using Gears

The chassis is the same as the one used in the fan version. The basic difference is in the way the wheels are fixed, and the gears transfer the power of the motor to the wheels. Figure 3.1.13 shows the chassis and the basic dimensions. See the cut in the part where the gears are installed.

First, prepare the chassis and find two pairs of wheels with the axis as shown in Figure 3.1.14. In one axis, place a plastic gear between 2 and 4 centimeters.

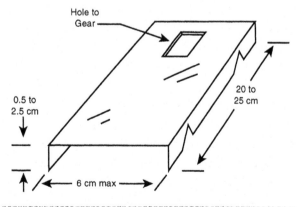

Figure 3.1.13 *The motor is coupled to the wheels through gears.*

The gear must be smaller than the diameter of the wheel in order to avoid contact with the ground.

In the motor's spindle, insert a small gear as shown in Figure 3.1.14. The two gears can be found in toys and electronic appliances. It is important to test combinations of gears that result in the best performance for the car. The performance changes according to the size of the wheels and the weight of the car.

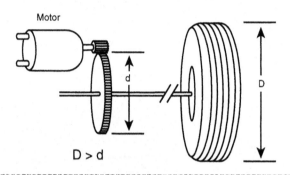

Figure 3.1.14 *The gear coupled to the wheels must be smaller than the wheels.*

The wheels with an axis are placed in the chassis as shown in Figure 3.1.15. See that the hole in the chassis allows the gear to adjust the motor without any contact with adjacent parts.

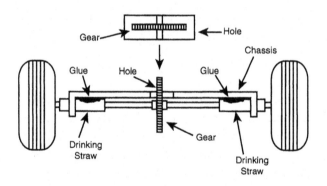

Figure 3.1.15 *The motor must be firmly coupled to the wheels' gear.*

Be sure that the wheels are aligned to permit the car to run in a straight line and reach the highest speed possible.

The next step in mounting is to place the motor. Use rubber bands as shown in Figure 3.1.16.

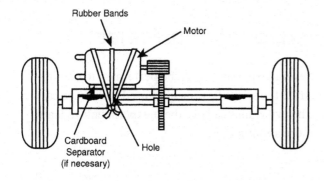

Figure 3.1.16 *Rubber bands are used to keep the motor in operation position.*

The motor can be placed in contact with the gears in other ways, but the advantage of using the rubber bands is that they can act as dampers to absorb the impact of obstacles or irregularities in the lane or in the alignment of gears.

Test the transmission system that powers the circuit to verify that all the power is transmitted to the wheels. If the wheels run backward instead of forward, invert the wires to the motor.

Now we can install the electronic circuit and the batteries as shown in Figure 3.1.17.

Protect the sensor from light when installing the circuit. A small piece of plastic or cardboard can be used. A pen cover can be used to execute this task. After wiring the circuit power, verify that the car moves in the correct direction. If it runs backward, invert the wires to the motor.

Testing the Car

1. Cover the LDR and insert the cells in the cell holder. Take care to avoid inverting any of the cells.

2. Align the car with others in the lane and remove the cover of the LDR. If the ambient light is sufficient, the motor will start and propel the car forward. If not, use a flashlight to illuminate the LDR.

3. If the motor does not run, check the contacts of the cells in the holder.

4. If the motor runs backward, invert the wires.

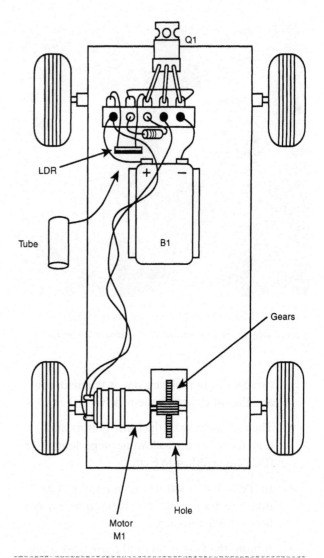

Figure 3.1.17 *The race car ready to be tested.*

Now the car is ready for the race. Start your engines and go!

The Race

Although the reader is free to create his or her own rules for a competition, I am experienced in facilitating this type of race. I have been organizing the event for the past 4 years in the college where I teach mechatronics. Figure 3.1.18 shows one of these competitions.

I have adopted the following rules:

- The pupils form teams of two or four members building one or two cars. In the case of

Figure 3.1.18 *The starting line seconds before a race, Colegio Mater Amabilis, Guarulhos, Brazil.*

two cars, as in real cars, one is the main car and the other acts as the reserve.

- The cars must have the same motors and dimensions in the following range: length, 15 to 25 cm; width, 4 to 6 cm.

- In the case of a school project, two evaluations are made: one for construction and one for race performance.

- If more than eight cars will participate, they can be divided into groups of two or three for the final race.

- The racer (student) can't touch the vehicle. A loss of points can be used to punish for such an infraction.

- The points attributed to the cars can be as follows:

Winner: 4 points
Second place: 3 points
Third place: 2 points
Cars that cross the final line: 1 point
Cars that complete half of the course: 0.5 point
Other conditions: combined by the organizer

- When using a flashlight as a remote control, the flashlight must be placed at least 30 cm from the LDR.

- Cars who crash or take other cars out of the competition due to changes in direction will be punished or disqualified.

Exploring the Project

Several experiments can be performed to demonstrate the scientific principles used in the race car. The evil genius can use these experiments to learn more about physics.

Newton's Laws

A domestic fan can be used to show the action and reaction principle. The force delivered by a fan to push the air causes an opposite force that simultaneously moves the fan. You can prove this by use of the following experiment.

Materials

1 domestic fan

1 skate

The Experiment

Demonstrate that when the fan pushes the air in one direction, the fan on the skate tends to move in the opposite direction, as shown in Figure 3.1.19.

How Gears Work

The reader can explain how gears work, calculating *the mechanical advantage* (TMA) of a pair of gears as a function of the number of teeth it has and its diameter, as shown in Figure 3.1.20.

Materials

2 small gears found in toys or other appliances (they must have different sizes)

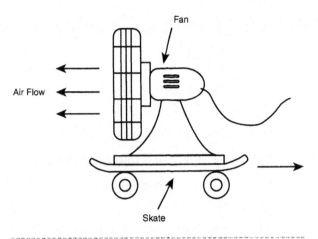

Figure 3.1.19 *Experiment to show how a race car that uses a fan works.*

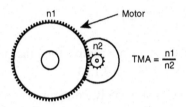

Figure 3.1.20 *Two plastic gears are coupled to study how torque and speed are changed.*

The Experiment

Show that with each turn of the larger gear, the smaller one will move a determinate number of turns. This number is calculated by the ratio between the numbers of teeth in each of the two gears. The power (torque) increases in the same ratio.

Cross Themes

Car, engines, fans, and movement are words often associated with the physics taught in grade schools. The following items can be used by teachers when exploring this project:

- **Analysis of movement** Analyze the movement of the race car. Is it a uniform or an accelerated movement? Explain the difference. Why does the race car stop accelerating after reaching a certain speed?

- **Friction** What is friction and how does it act on the movement of the race car? What can you do to reduce friction and increase the final speed of the car?

- **Calculating gears** What happens if the gears are changed to modify the motor's transmission ratio? How can you calculate this?

- **Fans** How do the number of fan blades determine the power of the car? What happens if you change the angles of the blades? What is the ideal angle?

Additional Circuits and Ideas

The basic idea of this project is to control a small DC motor using light. The sensor is an LDR and the "remote control" is a flashlight. However, the evil genius can upgrade the project with the following ideas for new circuits.

Direct Drive

A very simple version of the race car can be built without a remote control electronic circuit. The builder directly connects the cell holder to the motor as shown in Figure 3.1.21.

It is enough to insert the cells in the holder and let the car run. The rules for the race must be changed: Keep the car stalled by holding it in place with your hands, releasing it only at the start signal.

Using a Power MOSFET

The Darlington transistor can be replaced by a power *metal-oxide semiconductor field-effect transistor* (MOSFET) as shown in Figure 3.1.22. You only have to reduce the resistor to 100 kΩ.

Any common Power MOSFET rated to 2 amps or more (IRF720, IRF640, etc.) is suitable for the task.

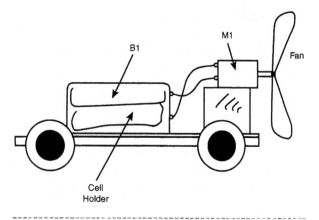

Figure 3.1.21 *Direct drive circuit. No electronic parts are needed.*

Parts List—Circuit Using a Power MOSFET

Q1   IRF640, IRF720, or equivalent Power MOSFET

LDR Common light-dependent resistor

M   DC motor of 6 volts or up to 500 mA

R1   100 kΩ × 1/8-watt resistor (brown, black, yellow)

B1   6 volts of power (4 AA cells and holder)

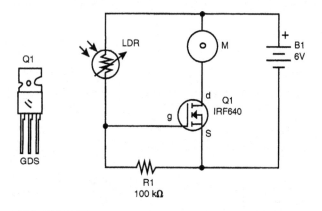

Figure 3.1.22 *Circuit using a Power MOSFET.*

Timed Circuit

Another interesting alternative for a race uses the circuit shown in Figure 3.1.23.

This circuit has a 555 IC configured as a timer. When the sensor (a reed switch) is triggered by a small magnet, the circuit turns on and the motor is powered by a time interval determined by C1 and the adjustment of P1.

When triggered, the race car will run for a certain time interval. P1 can be adjusted in order to make this interval long enough to cover the distance man-

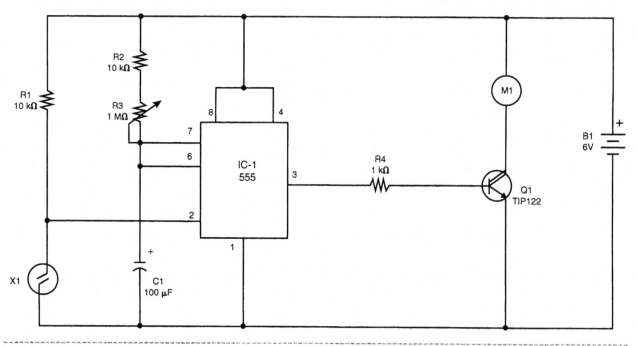

Figure 3.1.23 *Circuit for a timed control for the motor.*

dated by the competition. In a race, each competitor will use a small magnet to start their cars.

Parts List—Timed Circuit

IC-1	555 IC timer
Q1	TIP122 or equivalent, NPN Darlington transistor
X1	Reed switch
R1, R2	10 kΩ × 1/8-watt resistor (brown, black, orange)
R3	1 MΩ trimmer potentiometer
R4	1 kΩ × 1/8-watt resistor (brown, black, red)
C1	100 µF × 12-volt or 16-volt electrolitic capacitor
B1	6 volts of power (4 AA cells and holder)

Running on Rails

One problem observed in this race car is that the direction of movements depends on the precision of wheel alignment. In a race, cars will typically run in different directions, sometimes causing crashes or circular paths.

An interesting option to a race, one better used with gear-constructed race cars, is to make the cars run on rails as shown in Figure 3.1.24. Small, free wheels run over a cable used as a rail.

The Technology Today

Today, fan-propelled vehicles are used in swamps, but the same principle can be observed in airplanes and

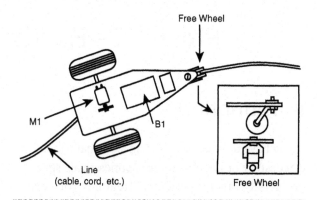

Figure 3.1.24 *Rails can be used to keep the cars in a straight (or curved) trajectory.*

helicopters. The transmission of power by gears is found in many vehicles and home appliances.

The power of your car's engine is transmitted by gears to the wheels, moving it at different speeds. The speed is determined by the ratio of the gears' teeth in the gearboxes. Many home appliances that use electric motors use plastic or metal gears to induce movement in mobile parts. Your electric mixer is one example.

Ideas to Explore

To get better performance or to learn more about the project, do the following:

- Try to use gears that reduce the speed of the fan and increase the power.
- Use a tube to direct the air of the fan as a turbine.
- Change the gears by increasing or reducing the diameter, noting which combination results in more power or more speed.

Project 2—RobCom: A Combat Robot

Introduction

If you have dreamed of creating a combat robot that can destroy or knock out the enemies, rivals, space invaders, or other supernatural creatures, you must take a look at this project. RobCom is the answer for your aspirations—your dream coming into reality. Using inexpensive and common parts, even inexperienced readers who may not yet be prepared to build complex circuits with microprocessors can build the RobCom.

The evil genius will be able to build the RobCom and challenge his or her friends to a real combat of robots. You can also invite your colleagues to build combat robots. Ask them to put into the project all their imagination, to create weapons, and to build defenses so they won't be knocked out in the first encounter.

RobCom was one of the projects proposed by the author to his pupils of the mechatronics course at Colegio Mater Amabilis in Guarulhos, Brazil. Figure 3.2.1 shows some of the projects made by the students. You can see how these evil geniuses constructed their destruction machines.

Figure 3.2.1 *Close-up showing two robots in combat.*

What Is RobCom?

RobCom is a remote-controlled robot built with common parts. In the basic version, it is built to carry a rubber balloon that it is protecting and three needles as weapons. The reader can add other weapons, of course, depending on the rules of combat agreed upon by the other competitors.

To make the project easier, the remote control uses a cable. Some advantages exist to using a cable in place of other remote control means, such as *infrared* (IR) or *radio frequency* (RF). Beyond the simplicity of the cable method (no special circuits are required), the problems of interferences and noise, common in the places where the combat takes place, don't exist.

RobCom has two small DC motors directly driving two rear wheels, which are made of standard CDs. The single front wheel is able to turn freely in any direction. The recommended front wheel is one that can be found in old office chairs or other furniture. Figure 3.2.2 shows the wheel used in the robot.

Figure 3.2.2 *Front wheel used in the RobCom.*

The control unit, which is placed at the end of a cable, is simply a small box with two special switches and a joystick. The switches can control two circuits (two poles) at the same time, and each switch has three positions. When the joystick is in the central position, the controlled circuit is disabled; no power is released to the motors. Certain positions of the switches control certain functions of the motor, as shown here:

Position	Motor
Forward	Motor runs forward
Released	Motor off
Backward	Motor runs backward

Combining the three positions of the switches, the robot can run in any direction, as the following table shows. The arrows indicate the direction of movement.

Switch A	Switch B	Symbol	Robot Movement
Released	Released		Stalled
Pressed forward	Pressed forward	↑	Run forward in straight line
Pressed backward	Pressed backward	↓	Run backward in straight line
Pressed forward	Released	↱	Turn right forward
Pressed backward	Released	↳	Turn right backward
Released	Pressed forward	↰	Turn left forward
Released	Pressed backward	↲	Turn left backward

RobCom is powered by four AA cells placed in the control unit. This placement reduces the weight of the mobile unit, increases the mobility, and is very important in combat.

Objectives—The Combat

The basic idea of robot combat is to put two Rob-Coms in an arena formed by four pieces of wood, as shown in Figure 3.2.3, and let them try to pop the other's balloon using needles as weapons.

Controlling the forward and backward movement of the robot, the player can find the best position for an attack without exposing his or her own balloon to the attack of the enemy. Strategies must be created by the RobCom evil genius to win the combat.

The winner is the first robot to pop the enemy's balloon. The balloons can have a tiny amount of flour or talc injected in them, adding a realistic effect for

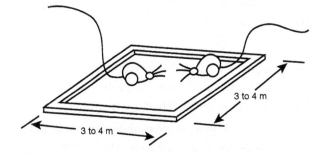

Figure 3.2.3 *The arena used in the combat is a square made by four pieces of wood.*

the explosion. A cloud of smoke will then announce that the balloon is popped and the combat is over.

A normal combat session usually lasts from 1 to 5 minutes (see Figure 3.2.3). In an organized competition, a simple elimination process can be used.

Rules are very important in order to avoid major differences among the competitors and to level the playing field. Rules include specifications for maximum dimensions of the robots, the size of the rubber balloon, the use of defense screens, the length of the needles, the types of motors used, the power supply voltage, and so on.

Rules for the actual combat are necessary, too. At the end of this project is a list of rules that we have found are useful in our combat.

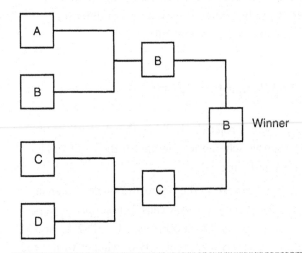

Figure 3.2.4 *Scene of combat where two RobComs test their power.*

Section Three   The Projects

The Project

The RobCom is a robot that runs on three wheels. The front two are used to turn the robot in the desired direction, and the third is a free wheel.

Two small DC motors are coupled directly to the wheels, which are made with CDs or some other material of the reader's choice.

The motors are controlled by a remote control, wired to the robot by a 3-meter cable. As indicated previously, the use of a cable makes the project easy to build and very inexpensive, as no special materials are needed. The remote control also houses the cells that power the motors. Two switches allow the robot to move backward and forward and to change direction.

The chassis can be made using common materials such as cardboard, CD boxes, plastic, wood, and so on. It is up to the reader to use his or her imagination to create his or her own version while following the rules of the competition.

When creating your RobCom, it is important to reduce the weight, making it as fast and well balanced as possible.

Building the RobCom

The basic RobCom is the one described in the introduction. Of course, the evil genius can change the RobCom by adding new weapons or defenses according to the combat rules.

The Electric Circuit

Begin your project by building the electric circuit. The simple schematic diagram for the RobCom is shown in Figure 3.2.5.

As we can see in the figure, the single power supply is formed by four AA cells. This supply powers two small, 6-volt DC motors via S1 and S2. S1 and S2 are a special arrangement of two switches with three positions each. In the middle position, the switches are off.

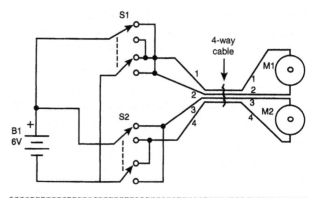

Figure 3.2.5 *Schematic diagram for the electric circuit of the RobCom.*

S1 and S2 determine the direction of the motors. The switches send the power to the motors by a special four-wire cable. The recommended length of the 4×26 *American wire gauge* (AWG) cable is 3 meters.

The switches are placed in a small box (plastic or other material) forming the joystick. The different colors of the cable wires are important and will help the reader know where each must be soldered. Figure 3.2.6 shows the electric circuit assembled.

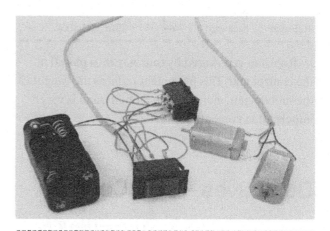

Figure 3.2.6 *The electric circuit ready to be installed in the RobCom.*

Parts List—The Electric Circuit

S1, S2	2 poles × 3 through switches (see text)
B1	6 volts of power (4 AA cells with holder)
M1, M2	6-volt DC motors

3 meters of four-wire cable (4 × 26 AWG)

Plastic box

Solder

The Mechanical Part

Figure 3.2.7 shows the basic mounting of the robot, detailing how the motor is coupled to the wheel. Figure 3.2.8 shows a rear view of the robot.

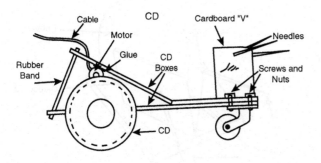

Figure 3.2.7 *Side view of the RobCom showing how the motor is coupled to the CD.*

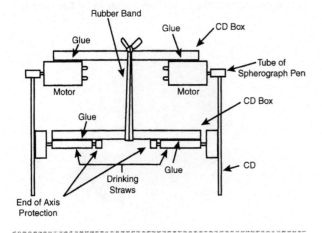

Figure 3.2.8 *The motors are glued to the CD box. They are kept in contact with the CD by the force of a rubber band. This makes it possible to transmit all the power to the robot.*

Putting the Pieces Together

The next figures show the sequence of operations to mount the robot. Figure 3.2.9 shows the free wheel fixed to the CD box used as a chassis and how to

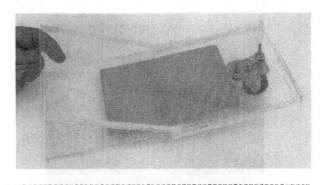

Figure 3.2.9 *Inserting a metal sheet between the CD boxes before gluing them together.*

insert a metal sheet between them to keep the structure rigid.

Figure 3.2.10 shows how to add an isolation sheet to the CDs to increase the adherence. Rubber bands glued to the CD can also be used to accomplish this. The reader is free to create the best way to increase the adherence, making the robot faster and agile.

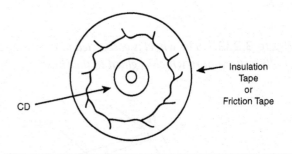

Figure 3.2.10 *Placing an isolation sheet to the CD to increase the adherence.*

The plastic wheel, taken from a toy, is glued to the CD. Small plastic cars and other toys are good sources for this wheel. I prefer the type with metallic axles. Figure 3.2.11 shows how to glue the wheel.

Pieces of cardboard can be used to make supports for the wheels. The spindles of the wheels are inserted into the drinking straws. At the end of the spindle, a small piece of a plastic cover connects the spindle to the cardboard support, as shown in Figure 3.2.12. This cover can be a small piece of the tube from a ballpoint pen or even the casing from an electric wire.

The motors are glued to the box. Be sure that the motors' shafts will be aligned with the wheels (the

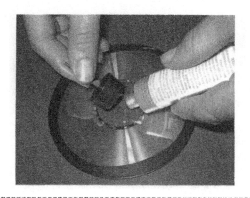

Figure 3.2.11 *Gluing the wheel to the CD.*

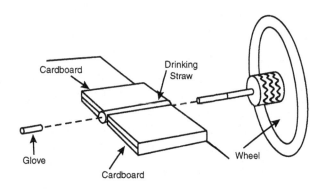

Figure 3.2.12 *The wheel is connected to a cardboard support, which is glued to the chassis.*

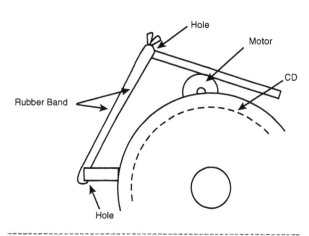

Figure 3.2.13 *Motors in place.*

CDs). Figure 3.2.13 shows the motors glued and contacting the CDs. The motors are forced to maintain contact with the CDs by a rubber band holding them in place.

To increase the transmission power of the motor to the CD, the spindle is covered with a small glove. As mentioned previously, the glove can be made using a piece of plastic tube taken from a ballpoint pen or even a plastic cover from an electric wire.

The arms of the robot are made with needles, placed in a piece of cardboard as shown in Figure 3.2.14. Figure 3.2.15 shows the RobCom ready for combat.

Finally, you can attach the rubber balloon to the robot using a rubber band.

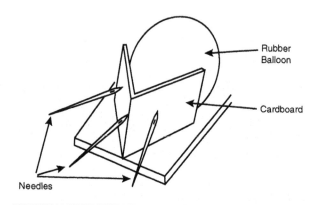

Figure 3.2.14 *The arms are placed in the front of the RobCom. This piece of cardboard also serves as a screen, protecting against the attacks of the enemies.*

Figure 3.2.15 *RobCom ready for combat.*

Testing the Combat Robot

Insert the batteries in the cell holder. Pressing the switches in the control unit, the motors should be activated. If not, check the solder and the cable. If one or both motors run in the opposite directions (e.g., forward when you press backward), invert the wire of the motor.

With the robot on the ground, test to see if the robot will move freely in all directions when you press the controls. If the motors have difficulties in moving the robot, verify that they are pressing against the CDs with the necessary force. If all the movements are satisfactory, your RobCom is ready for combat.

The Combat

The reader is certainly free to create his or her own rules for competition. However, we can offer some suggestions based on the experience of many combat sessions organized in the school where the author teaches. Figure 3.2.16 shows several RobComs in an arena, waiting for the beginning of the contest.

Combat Rules and Specifications for the Robots

To avoid major differences among the robots, it is important to establish some rules regarding the robot and the competition. The robot characteristics should be as follows:

- The length of the robot must be between 15 and 25 centimeters.
- The maximum number of needles used as arms should be three.
- The maximum length of the needles should be 20 centimeters (including the support).
- The screen in front of the robot should be limited to 10×15 centimeters.
- All the robots should use the same type of motor.
- The power supply must be four AA cells for all robots.
- The rubber balloons must be equal in size.
- No other weapon is allowed (or otherwise combined).
- The arena is formed by four pieces of wood with the dimensions between 3×3 and 4×4 meters.

Combat Rules

- The competitors cannot enter the arena.
- The competitors cannot pull the robots by the remote-control cable.
- The competitors must begin at opposite corners of the arena.
- The combat begins when the referee gives the order.
- The combat ends when the balloon of one competitor is popped.
- If the two balloons are exploded at the same time, the robots must compete against each other in a second round to determine the winner.

Figure 3.2.16 *Combat of RobComs at Colegio Mater Amabilis in Guarulhos, Brazil.*

Exploring the Project

Many changes can be made to the original project specifications. The following are simple examples of the changes that can be made:

- Wood rods can be used to substitute for the plastic chassis.
- A plastic chassis with different formats can be used.

The same project can also be used with vehicles other than combat robots. The reader can mount his or her own walking robot using the control system described here. Figure 3.2.17 shows such a robot.

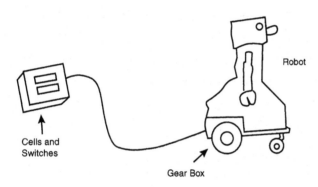

Figure 3.2.17 *A common robot using the remote control described in this project.*

Cross Themes

The control of movement is a theme that translates easily into the physics curriculum. Teachers who want to explore this as a cross theme can invite the pupils to research the next items:

- **Analysis of movement**: Describe the type of movement the robot makes during combat and how fast changes in direction or speed can alter the equilibrium.
- **Friction**: Analyze the effect of friction in the robot mobility.
- **Transmission**: Find the best way to transmit the power from the motor to the wheels. Think about the use of gears.

Additional Circuits and Ideas

The basic circuit used in the project is very simple. It uses no electronic pieces and no complex elements. The reader who is experienced in electronics can upgrade the circuit and create some interesting projects.

Using a Joystick

Figure 3.2.18 shows how a common joystick, such as the ones found in video games or PCs, can be used to control the two motors in the RobCom.

The circuit uses four relays to control the two motors. The four switches of the joystick are used to power the motors on and off or to reverse the current across them. The position of the joystick affects what happens to the motor:

Joystick Position	Left Motor	Right Motor
Center	Locked	Locked
Up	Forward	Forward
Down	Backward	Backward
Right	Forward	Backward
Left	Backward	Forward
Upper right	Forward	Locked
Upper left	Locked	Forward
Lower right	Locked	Backward
Lower left	Backward	Locked

Parts List—Using a Joystick

D1 to D8   1N914 or equivalent silicon diodes

K1 to K4   6 V to 12 × 50 mA reversible relays (select voltage according the motor)

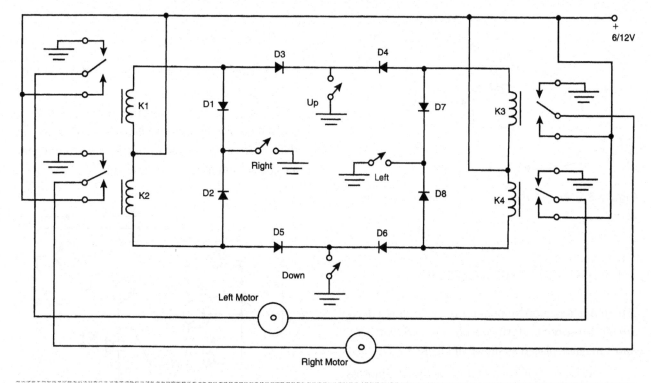

Figure 3.2.18 *Controlling the combat robot with a game joystick.*

Adding a Pulse Width Modulation (PWM) Control

A PWM control for the motor will allow the reader to change the speed of the RobCom. The PWM control is described in detail in Project 3.3.

Figure 3.2.19 shows how to add a PWM block to the RobCom. Notice that only one PWM control is necessary to change the speed of both motors. Or if the reader wants, he or she can use one PWM control for each motor.

Adding Weapons

The reader can put his or her imagination to work creating new weapons for the RobCom. Of course, it is important to be sure the weapons meet combat rules.

Figure 3.2.20 shows how a small DC motor can be used to add movement to the needle, making it much more dangerous as a tool against the enemy.

Another idea is to couple a rotary ball with needles, as shown in Figure 3.2.21. In this case, the reader must take care that the ball does not pop his or her own balloon.

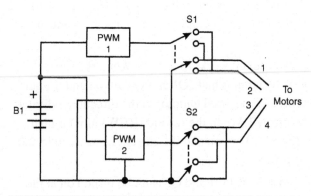

Figure 3.2.19 *A PWM control is added to the RobCom to control the speed.*

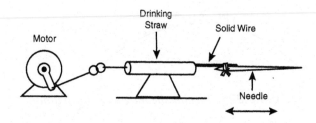

Figure 3.2.20 *The motor coupled to the needle can be activated by diodes when the RobCom runs forward.*

Section Three   The Projects

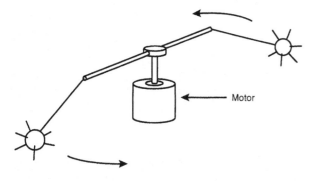

Figure 3.2.21 *A rotary weapon for the combat robot.*

Adding a Death Circuit

An interesting improvement for the project is a *death circuit*. This circuit is formed with two reed switches and a magnet, placed as shown in Figure 3.2.22.

As we can see, the current flows across the motors and passes through the reed switches. The magnet is attached with a rubber band inside the balloon. If the balloon is full, the magnet touches the reed switches, and the motors are powered.

If the balloon pops, the magnet falls and the reed switches open. The motors are no longer powered, and the robot stalls.

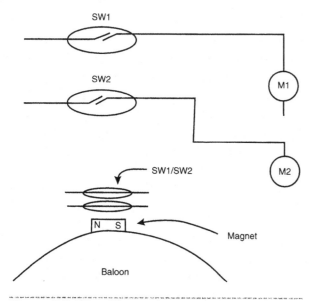

Figure 3.2.22 *Placing the magnet in the balloon.*

Adding Sound

Figure 3.2.23 shows a simple sound-effect circuit for the RobCom.

If a 47 nF capacitor is used, the circuit generates sounds like a siren produces when the motors are activated. If a 10 μF capacitor is used, the circuit generates pulses imitating a machine gun.

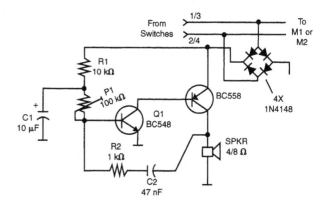

Figure 3.2.23 *Sound-effect circuit for RobCom.*

Using an H-Bridge

The digital control of the RobCom can be implemented by using an H-bridge. This idea is based on the fact that four transistors can be used to control the current across a motor in the same way a *double-pull/double-throw* (DPDT) switch controls the current. The circuit proposed in Figure 3.2.24 is a full bridge or H-bridge using four Darlington transistors.

Using an H-bridge for the control has two advantages. First, the current across the cable is reduced. Second, logic signals can be used in the control. In this case, even a computer can be used to control the RobCom.

The circuit works as follows: When the forward (FWR) input is high, Q1 and Q4 are on and the current flows across the motor in the direction indicated by arrow 1. When the rewind (REW) input is high, Q2 and Q3 are on and the current flows as indicated by arrow 2.

Notice that Q1 and Q3 can't be turned on at the same time, as can Q2 and Q4, because that would

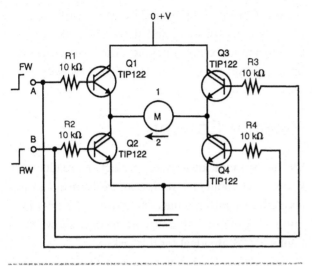

Figure 3.2.24 *The circuit for an H-bridge.*

Input A	Input B	Motor
Low	Low	Stalled
High	Low	Run forward
Low	High	Run backward
High	High	Forbidden

Figure 3.2.25 shows how to add a fifth transistor and a logic system to avoid this forbidden state.

Parts List—Using an H-Bridge

IC1	4011-4 NAND gates (CMOS *integrated circuits* [ICs])
Q1 to Q5	TIP122 *negative-positive-negative* (NPN) Darlington transistors
R1 to R5	10 kΩ × 1/8-watt resistor (brown, black, orange)
M	DC motor (up to 500 mA)

mean a short circuit for the current between +12 volts and ground. It is a forbidden state that can cause the transistor to burn. The circuit can be controlled in the following ways:

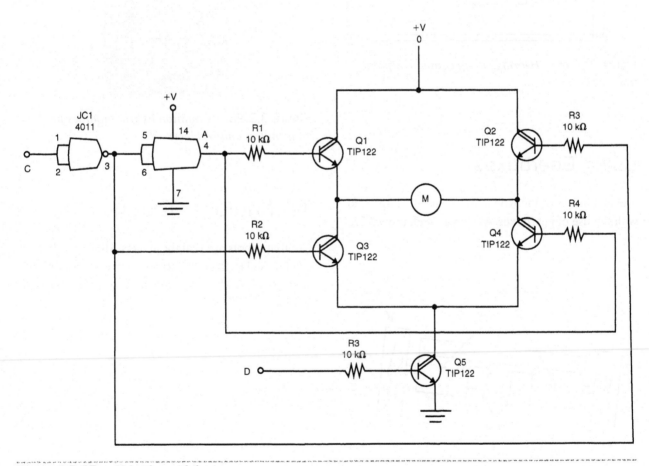

Figure 3.2.25 *H-bridge with logic.*

The bridge works in the following manner:

Input C	Input D	Motor
High	High	Run forward
Low	High	Run backward
x (don't care)	Low	Stalled

This same bridge can be implemented using the same bipolar transistors used in the BD135 (500 mA) or TIP31 (2 A), as shown in Figure 3.2.26.

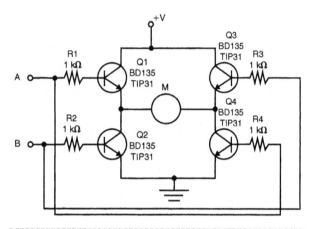

Figure 3.2.26 *H-bridge using common bipolar transistors.*

Using Gearboxes

The mechanics of the RobCom can also be improved with the use of gearboxes. As shown in Figure 3.2.27, the RobCom can be powered by small gearboxes, increasing the efficiency and allowing the robot to be built in a more compact manner. The gearbox can be used to power the CD wheels or plastic wheels.

Remote Control

Starting with the ideas in this project, the reader who knows a lot about electronics can easily install a wireless remote control in the combat robot. Small transmitter/receiver modules, such as the ones shown in Figure 3.2.28, are ideal for this task.

When using wireless remote controls, care should be taken by the builders to choose different frequencies for the systems. If two robots use the same frequency, interference problems will put the combat in jeopardy.

Figure 3.2.28 *Common hybrid modules for transmitters and receivers.*

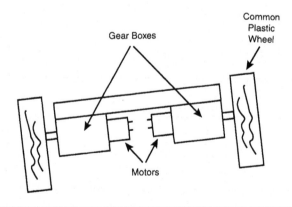

Figure 3.2.27 *Gearboxes can be used to power the robot. The photo shows a gearbox suitable for this project.*

Technology Today

Combat robots are real these days. Military robots can be used to fight, to carry resources, on rescue missions, or to go where it is too dangerous for a human soldier.

Project 3—Using a PWM Motor Control

Introduction

Pulse width modulation (PWM) motor controls are very efficient for controlling the speed and torque of small DC motors. Adding a precise control to a DC motor, the evil genius can develop many interesting projects involving remotely controlled movement.

The basic version of the project described here can be used to control small 3- to 12-volt DC motors, draining currents up to 500 mA. Some improvements in the basic project can extend both the range of currents and the voltage of the motors.

Objectives

The main objective of this project is to give the evil genius the resources to control small motors in mechatronics devices. The following is a short list of suggested projects that can use a PWM control like the one described here:

- Automatic elevator
- Automatic window
- Speaking head
- Fan
- Model car

Of course, using his or her imagination, the evil genius can put the PWM control described here to work in many other projects.

The Project—How It Works

There are two ways to control small DC motors. The simplest way is to control the power applied to a load by using a rheostat wired in series with the motor (see Figure 3.3.1). This configuration is called a linear control.

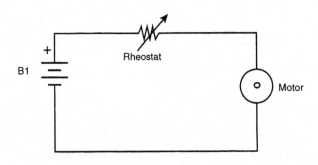

Figure 3.3.1 *The simplest way to control the power applied to a load is to wire a variable resistor in series.*

Together the rheostat and the load form a variable voltage divider. By changing the resistance of the rheostat, dhe voltage applied to the motor also changes and consequently so does its speed and torque.

The main inconvenience in this type of control is that the current pulled, or used, by the motor passes across the rheostat. Therefore, the greater the amount of current pulled by the motor, the greater the amount of heat generated in the rheostat. Another problem to consider is that, because the motor represents a variable load, the circuit involved in controlling the DC motor can be unstable, mainly at low speeds. It is very difficult to make the motor start softly while keeping the torque constant along the range of operation.

Motors are variable loads because they change their electrical characteristics as the speed and torque changes. In a practical linear control, the rheostat is not used to directly control the current across the motor, but the current through the base of a transistor, as shown in Figure 3.3.2.

So the rheostat dissipates less heat as the transistor is used to control the main current in the circuit, and therefore the transistor dissipates more heat. In this book we will give details of how to construct a linear power control, also known as an electronic rheostat.

Section Three The Projects 33

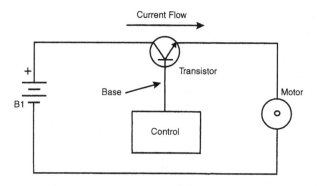

Figure 3.3.2   *The main current across the motor is controlled by a transistor.*

The second way to construct a linear power control is by using the PWM control. Because this is a much more efficient way to control the amount of power applied to a load, it is also used in other applications' power supplies, such as common appliances like computers, TVs, VCRs, and so on. A PWM control is also called a *switched mode power supply* (SMPS).

The basic idea of a PWM starts from the use of square voltage pulses that power a motor (see Figure 3.3.3).

The amount of power applied to the motor depends on the duration of each pulse, or the duty cycle, of the control voltage. If the duration of the pulse is the same as the interval between the pulses, we say that the duty cycle is 50 percent, and the average power applied to the load is 50 percent (see Figure 3.3.3a).

If the duration of the pulses is extended, the average power applied to the motor increases by the same proportion. Figure 3.3.3b shows what happens in this scenario.

It is easy to see that controlling the width of the pulses can change the average power applied to a motor or to any other load. The process used to control the width of the pulses that are applied to the load is called *modulation*, and this kind of circuit is termed pulse width modulation (PWM) power control.

With this in mind, we can move on to a practical PWM control. A basic configuration for a practical PWM is given in Figure 3.3.4.

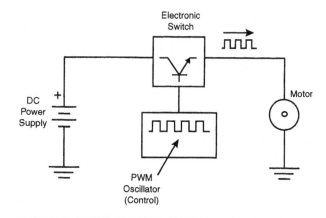

Figure 3.3.4   *A transistor is used as a switch to turn the current across the motor on and off.*

A bipolar transistor (or other transistor) is wired to a variable-duty-cycle oscillator. When the oscillator is running, the transistor turns on and off at the frequency of the oscillations, applying a square wave to the motor. Note that, when the transistor is on, the resistance between the emitter and the collector can be considered to be zero and no power is generated across it. (The power is the product of the voltage drop and the current; because the voltage is practically zero, the result is zero power.) When the transistor is off, no current flows through it and the power dissipated is zero, too.

In the real world, the transistor can't pass from the on state to the off state, or vice versa, quickly enough to avoid the generation of heat. The transistor needs a finite time for the change of state. It is important to note the heat generated by the switching process in this kind of circuit. However, it is generally very low as compared to the heat generated by linear controls.

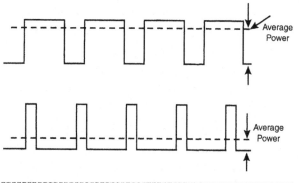

Figure 3.3.3   *Operation principle of a PWM control.*

Another advantage of PWM controls happens when controlling DC motors. DC motors controlled by PWM can maintain more constant torque over the entire speed range.

The Basic Project

A PWM control for small DC motors can be implemented in a variety of ways using common parts. Our basic project uses a common 555 *integrated circuit* (IC) in a stable version, driving a *positive-negative-positive* (PNP), medium-power transistor, as shown by the blocks in Figure 3.3.5.

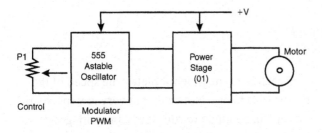

Figure 3.3.5 *Basic PWM control.*

This circuit is not a real PWM but a quasi-PWM. Using fewer components, we can build a simple configuration that functions on a basic level and is enough for our purposes. At the end of the project, we will give other versions for the reader who wants to do more with this kind of circuit.

This circuit is formed by a 555 timer IC operating in a stable configuration. The circuit generates a square wave, with the frequency changing by the change of the duration of the pulses, as shown in Figure 3.3.6.

The result is that although the average frequency changes (which doesn't happen in a real PWM control), the pulse width changes, too. This is why we say it is a quasi-PWM control. Therefore, it is possible to control the speed and torque of the motor in a range from almost 0 percent to 90 percent of the maximum.

The central frequency is determined by the C1 capacitor. The builder must choose a capacitor that matches the characteristics of the motor. Low-value capacitors can produce very high frequencies, and the

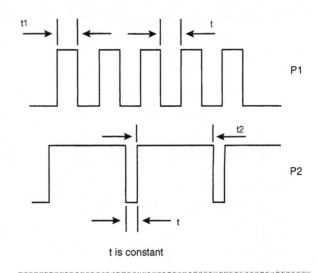

Figure 3.3.6 *In our circuit both the pulse width and the frequency change.*

motor won't react to the signals. The efficiency of the circuit will fall in this case, and even the motor cannot start.

On the other hand, capacitors with very low values can cause the motor to vibrate or run with small jumps at low speeds. The reader must experiment to find the best capacitor in the range of values suggested in the text, matching the characteristics of the motor being used.

The circuit can operate from power supplies ranging from 5 to 12 volts, and the motor can pull a current up to 500 mA. Later in the chapter, in the section on upgrades, we suggest circuits that can drive more powerful motors.

How to Build the PWM Control

Figure 3.3.7 shows the schematic diagram for the PWM control using the 555 IC.

The reader can mount the circuit in a small *printed circuit board* (PCB) with the pattern suggested in Figure 3.3.8.

Of course, the evil genius is also free to make his or her prototype using protoboards or universal PCBs.

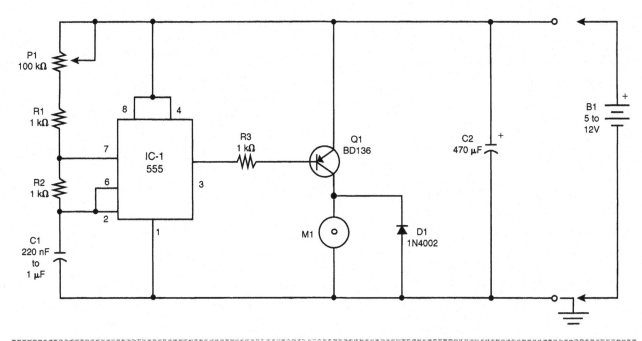

Figure 3.3.7 *Schematic diagram for the basic PWM control.*

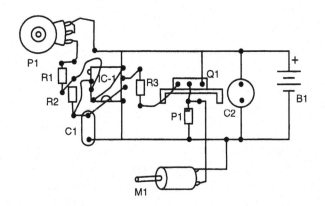

Figure 3.3.8 *Small PCB holding the electronic components for the PWM control.*

The transistor must be mounted on a heatsink, and the potentiometer can be wired to the circuit using long wires. If the motor must be placed far from the control unit, it is better to mount the circuit near the motor and not near the potentiometer.

Test and Use

Connect the control to a DC motor and to a power supply. The power supply can be formed by using four AA cells or one D cell (for a total of 6 volts).

Reacting to the potentiometer, the speed of the motor will change. Notice that at very low speeds, the motor will continue to run slowly, but with good torque.

If the motor tends to vibrate at low speeds, or not to reach the maximum speed, change the value of the C1 capacitor until you find the best value for the desired performance.

Parts List—PWM Control

IC	1-555 IC timer
Q1	BD136 or equivalent (BD138, TIP32, TIP42, etc.) medium-power PNP transistor
D1	1N4002 silicon diode
R1, R2, R3	1 kΩ (brown, black, red)
P1	100 kΩ linear potentiometer (rotary or slide)
C1	220 nF to 1 μF polyester or electrolytic
C2	470 μF × 16 *working voltage in direct current* (WVDC) electrolytic
B1	5- to 12-volt power supply

M1   DC motor, 5 to 12 volts, up to 500 mA

PCB, heatsink for Q1, wires, solder, etc.

Projects

Many interesting projects can be built using this control. You can even add this control to motors in other projects described in this book, such as the RobCom.

Simple Elevator

Figure 3.3.9 shows how to mount an experimental elevator using this circuit. S1 is a switch that changes the direction of the motor.

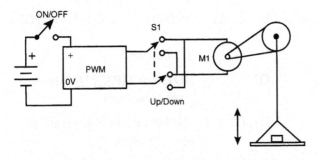

Figure 3.3.9 *Simple elevator using the PWM control.*

The gearbox can be found in toys, electronic appliances, or other mechatronic devices. You can also reduce the speed and increase the torque of the system by using a reduction system made with rubber bands (see Figure 3.3.10).

Automatic Window

Figure 3.3.11 shows how you can use a small DC motor with a gearbox and PWM control to make an automatic open and close system for your window.

It is important to prevent bumps at one end of the course or the other (when the window is totally open or closed) so no hitches occur when using switches to open the circuit.

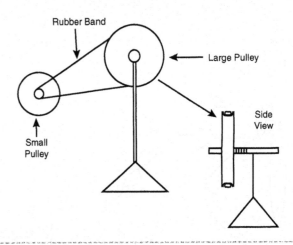

Figure 3.3.10 *Reducing the speed and increasing the torque of a motor with rubber bands.*

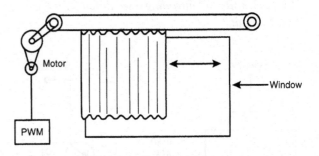

Figure 3.3.11 *Automatic window using the PWM control.*

Fan

Figure 3.3.12 shows how a small electric table fan can be built using a PWM control. The speed can be controlled across a large range of values, according to the amount of breeze you want.

Increasing the Power

Figure 3.3.13 shows how a Darlington transistor can be used to control high-power motors.

A TIP122 can be used to power motors up to 2 amps. The transistor must be mounted on a heatsink. Other transistors such as the TIP140 can be used to control motors up to 4 amps.

A Power MOSFET can be used in the circuit without causing any visible change. Remember that Power MOSFETs need 9 volts or more for them to have good control of a load.

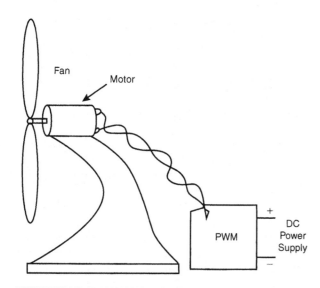

Figure 3.3.12 *Controlling a small fan with the PWM circuit.*

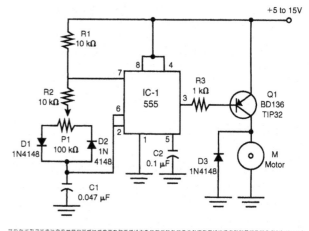

Figure 3.3.14 *Schematic diagram for the PWM control.*

BD136 or a Darlington, if you want to control powerful motors.

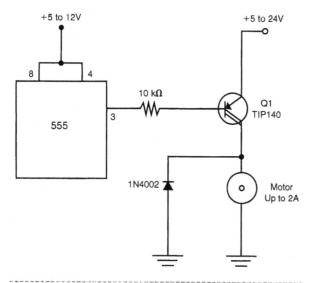

Figure 3.3.13 *Using a Darlington transistor to increase power.*

A Real PWM Control

The circuit shown in Figure 3.3.14 uses two diodes in an arrangement that changes at the same time a change occurs in the pulse width and in the interval between pulses. By using this circuit, it is possible to keep the frequency almost constant even through variations occur in the duty cycle.

This configuration is better than the one we had suggested as the basic project. The output transistor can be a medium-power PNP type, such as the

Parts List—Real PWM Control

IC1	555 timer IC
Q1	BD136 or TIP32 PNP silicon power transistor
D1, D2, D3	1N4148 or 1N914 general-purpose silicon diodes
R1, R2	10 kΩ × 1/8-watt resistor (brown, black, orange)
R3	1 kΩ × 1/8-watt resistor (brown, black, red)
P1	100 kΩ potentiometer
C1	0.047 to 0.47 μF capacitor
C2	0.1 μF capacitor

Controlled by Light

Another interesting idea is to build a robot that can follow the light (or avoid a light) source. The potentiometers are replaced by *light-dependent resistors* (LDRs) and mounted in an arrangement like the one shown in Figure 3.3.15.

The amount of light falling onto the sensor determines the speed of each motor and then the direction

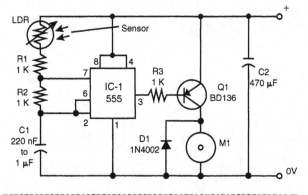

Figure 3.3.15 *By replacing the potentiometer by an LDR, the speed is controlled by the light falling on the sensor.*

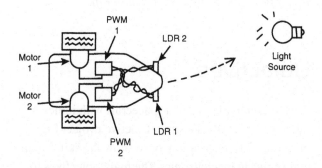

Figure 3.3.16 *Movement of a robot following a light source.*

of the movement of the robot. Depending on the way the sensors are placed, the robot will follow or avoid the light, as shown in Figure 3.3.16.

To get more directivity and sensitivity for the sensor, you can mount them onto opaque cardboard having a convergent lens in the front, as shown in Figure 3.3.17.

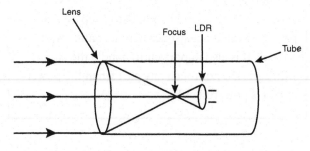

Figure 3.3.17 *Installing the LDR inside a cardboard tube with a convergent lens to increase the sensitivity and directivity.*

Controlling with a Computer

The circuit shown in Figure 3.3.18 can be used to convert the logic level in the *input/output* (I/O) port of a computer to a voltage. This *digital-to-analog converter* (DAC) can be used to control the speed of a DC motor when coupled to the PWM control.

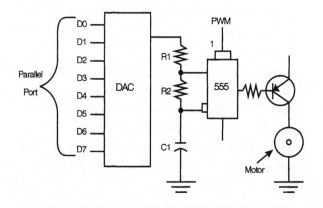

Figure 3.3.18 *A DAC can be used to control the speed of a motor with the help of a computer.*

The Technology Today

PWM controls are used in industrial machines to keep the speed under predetermined values. The precision of the control and the use of feedback sensors to monitor the speed make this kind of circuit ideal for those applications.

Section Three   The Projects

Project 4—Ionic Motor

Experiments and projects using high voltage have always been fascinating. The main objective of the project described here is to build a fantastic engine, called an *ionic motor*. Such engines are now used in low-power operations in space and probably will power large spacecraft in the future. Many other projects and experiments can be performed using this same basic circuit.

Ionic motors are used in spacecraft outside the earth's atmosphere because they can deliver enough power to be useful in a vacuum. They are the fastest engines that man can build. The flux of ions released by an ionic motor can accelerate a spacecraft to theoretical speeds up to 80,000 *kilometers per second* (km/s).

Unfortunately, practical ionic motors are very weak and need many weeks or even years to accelerate a spaceship to high speeds. The thrust of a large ionic motor isn't more than a few grams.

NASA is working hard to create powerful ionic motors, but it is a slow process of creation and experimentation. They will need many more years of experimentation to have practical motors with power enough to move a large spacecraft.

Currently, ionic motors are used to change the position of satellites in their orbits and to make small corrections in their trajectories.

What we propose here is an experimental ionic motor that you can build as a propeller for a miniature spacecraft. You can make a miniature of the *Star Trek Enterprise* or the *Star Wars'* pod racer with your operating ionic motor. Figure 3.4.1 shows a spacecraft made with a neon lamp that moves in a circular trajectory and is thrust by a flux of ions.

In this project, the evil genius will learn how to mount a high-voltage circuit to create the flux of ions in the engine and how to build the mechanical system for the spacecraft.

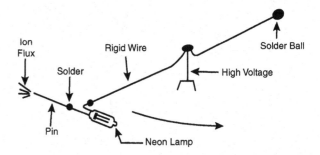

Figure 3.4.1 *Spacecraft built by the author with a small neon lamp. The lamp glows at the same time that a flux of ions thrusts the spacecraft.*

Objectives

Certainly, a miniature of a spacecraft with ionic motors that works is a very attractive project for science fairs, as a decoration, or to show the skills of an evil genius in mechatronics. The main objectives for this project are to accomplish the following:

- Show how an ionic motor functions.
- Show practical application of the "point effect."
- Make experiments with a *very high voltage* (VHV).
- Show how Newton's law also works in the space.

> **Caution!** This circuit operates with VHVs. All necessary care should be taken to keep all the parts protected against accidental contact with anyone. It is also important to handle the circuit with care. Do not touch any part when the power circuit is powered on!

Ionic Motors

Ionic motors operate according to a phenomenon known for many centuries: the *point effect*. Electric charges on a body (a solid conductor), because of the long-range action of the electric forces, have the tendency to accumulate on the borders of a conductive body. This is commonly known as the *skin effect*. If the body is spherical, as shown in Figure 3.4.2, the electric charges are distributed uniformly around its surface.

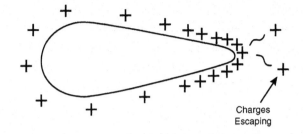

Figure 3.4.3 *The point effect.*

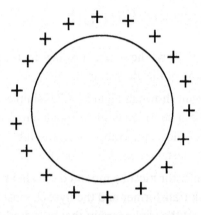

Figure 3.4.2 *The charges are uniformly distributed around the surface of a sphere.*

The electric pressure that moves the charges to the borders of the body depends on the density of the charge and on the electrical field. We can say that the electrical pressure is proportional to the square of the charge density. This means that the electrical pressure is nonlinear (if you double the field, the pressure will be multiplied by 4).

Combining the skin effect with the nonlinearity of the pressure, we can explain why the electric field is higher in sharp regions of a solid conductor, resulting in what we call the point effect. As Figure 3.4.3 shows, the charges tend to accumulate in the sharp regions of a conductor.

A related important phenomenon is part of this process and will be important for our application: *the breakdown*. If the charges in the sharp regions of a body reach a voltage that is high enough, a breakdown occurs. The air becomes conductive, and the charges can be expelled in a fluid that is moving. The science that studies this fluid in movement is called *electrofluid dynamics* (EHD).

But if the movement of the charges is strong enough to cause collisions in the particles that carry the charge (i.e., in the air), the temperature will rise, the electrons will be lost, and light will be produced. The air around the sharp part of the body becomes a plasma.

According the nature of gas, the energy level of the electrons is different, and thus a color of light is emitted. Neon produces orange light, and the air is yellow, red, or blue according to the energy level involved in the process.

The charges are transferred to the surrounding air particles and then repelled, providing a force that can be used to move the spaceship. According Newton's law, the force used to repel the charges results in an equal force in the opposite direction by which the spacecraft is moved (see Figure 3.4.4).

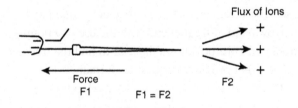

Figure 3.4.4 *Newton's law explains how the ionic motor functions.*

Although the power released in the process is very low, the speed generated by this kind of propulsion (by the charge particles) is fantastic: 80,000 kilometers per second!

Of course, when we combine very low power with very high speed, we must also be able to build spacecraft that can withstand years in space and fantastic speeds. A typical ionic spacecraft will need many years to reach the approximately 10,000 km/s (36,000,000 km/h) according to the specialists working to make them ideal to explore other solar systems.

How Our Project Works

Our basic circuit is made up of a high-power Hartley oscillator and one transistor. This circuit can deliver voltages up to 30 kV, enough to power the spacecraft. In a Hartley oscillator, a coil serves as both the load and the feedback network that keeps the circuit in oscillation (see Figure 3.4.5).

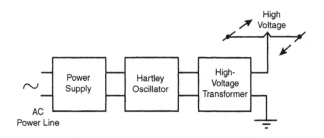

Figure 3.4.5 *Part of the energy released by the circuit is converted to the high voltage that is powering the spacecraft, and part is used as feedback to maintain the circuit in oscillation.*

The load in parallel with C3 forms a tuned circuit that, along with the network circuit, determines the oscillation frequency. The circuit can be tuned to a narrow band of frequencies with the use of a potentiometer. This tool allows the user to find the frequency that provides the best performance. The potentiometer actually acts as an accelerator.

The coil used as the load for the oscillator is also the primary coil of the high-voltage transformer. In the original project, we used a flyback transformer found in an old TV. Any flyback can be used. You simply have to take care to find one where part of the core is uncovered. In that part, we have to make the primary coil. Various flybacks can be found at Information Unlimited (www.amazing1.com/transformers.htm).

The transistor is a high-power *positive-negative-positive* (PNP) silicon transistor MJE15004. This transistor can handle the high voltages and currents delivered by the circuit.

The power supply consists of a transformer, two diodes, and a filter capacitor. The transformer reduces the power supply line to 12 volts. This voltage is then rectified by D1 and D2 and filtered by C1. The secondary purpose of the transformer is to supply at least 3 amps to support the power delivered from the circuit.

How to Build the Electronic Circuit

Figure 3.4.6 shows the schematic diagram for the basic version of the electronic circuit.

The circuit shown in Figure 3.4.7 is simple and uses parts that can be found easily through any dealer. Figure 3.4.7 shows the components assembled and ready to be installed inside a wooden box.

The transistor must be mounted on a heatsink. T2 is a flyback transformer and the flyback must be the type without the tripler circuit that uses diodes and capacitors. This kind of transformer is protected by a plastic cover to avoid any contact with the core. Therefore, it is impossible to add the primary coil required in the project.

The primary coil is made up of 8 + 8 or 12 + 12 turns of 20 to 26 common plastic-covered wire as shown in Figure 3.4.8.

P1 requires a wire-wound potentiometer because the current across the base of the transistor is high. The correct number of turns depends on the flyback. You will need to experiment to find the best value. R1 is a wire-wound resistor.

When testing the circuit, you might alter C2 and C3 to find the best performance. C2 can be a capacitor in the range of 0.01 μF and 0.1 μF, and C3 can be a capacitor in the range of 0.100 μF and 0.560 μF.

All the components fit into a wooden box as shown in Figure 3.4.9. The power transistor is placed on a heatsink, and the transformer is attached firmly in the box with screws. The box used in this project measures 22 × 22 × 15 centimeters. This box, purchased in an office supply store, is generally used to file cards.

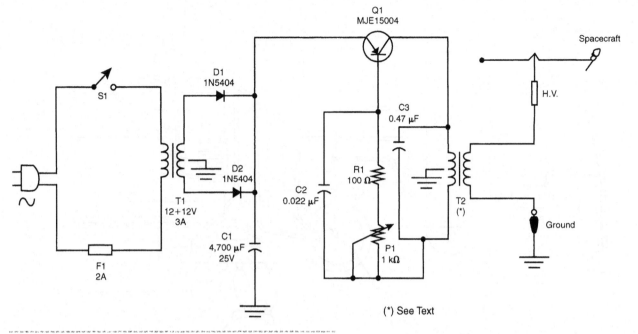

Figure 3.4.6 *Electronic circuit for the ionic motor.*

Figure 3.4.7 *The circuit on a workbench, assembled and ready to be tested.*

Figure 3.4.8 *Detail of the primary coil of T2.*

Figure 3.4.9 *The electronic circuit fits inside a wooden box.*

The flyback is also placed inside the box and kept firmly in place with screws. The high-voltage terminal passes through a hole in the cover of the box so that it can provide power to the spaceship. A 2-amp fuse protects the circuit against shorts or other problems.

When mounting, observe the position of the polarized components such as the transistor, diodes, and electrolytic capacitors. Any accidental change in position can burn the component or others in the same circuit. Keep all the wires short in length. Figure 3.4.10 shows the parts placement diagram.

Section Three   The Projects   43

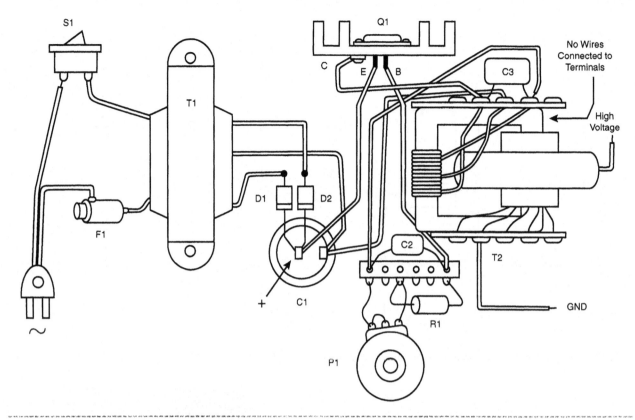

Figure 3.4.10 *Parts placement diagram; no printed circuit board (PCB) is necessary.*

Don't forget the ground wire, a piece of about 3 meters of common wire connected to one of the lateral high-voltage terminals of the flyback. To find the ideal terminal, the reader will need to experiment.

Testing the High-Voltage Circuit

After taking a good look at all the connections, verifying that each one is okay, you can turn on the power switch. A soft noise will reveal that the circuit is oscillating and a high voltage is being produced.

Place a fluorescent lamp (5 to 40 watts) near the high-voltage terminal and it will glow due to the presence of the high-voltage field. This is a good way to test the circuit.

Parts List—High-Voltage Circuit

Q1	MJE15004 high-power *negative-positive-negative* (NPN) silicon transistor
D1, D2	1N5404 silicon rectifier diodes
R1	100 Ω × 5-watt wire-wound resistor
P1	1 k Ω wire-wound potentiometer
C1	4,700 µF × 25-volt electrolytic capacitor
C2	0.022 µF polyester film or ceramic capacitor
C3	0.47 µF polyester film or ceramic capacitor
T1	Transformer, with primary coil rated according to the power supply line, and secondary coil rated to 12 + 12-volt × 3-amp *center tap* (CT)

T2	flyback transformer (see text)
S1	SPST switch (optional ganged to P1)
F1	2-amp fuse with holder

PCB, terminal strips, heatsink, screws, knob for P1, solder, power cord, wires, box, and so on

Figure 3.4.12 *The ionic propeller can be adapted to a miniature of the pod racer seen in* Star Wars.

The Spacecraft

A simpler and lighter-weight spacecraft can be made using a neon lamp, as shown in Figure 3.4.11.

Figure 3.4.11 *Spacecraft built with a neon lamp.*

The advantage to using a small neon lamp or a xenon flash lamp is that they both glow when the circuit is powered on and while the spacecraft moves in its circular trajectory.

One terminal of the lamp is soldered to the support and the other to a pin, as shown in the figure. It is very important that each place where solder is used is rounded and without any points. If a point exists in any part of the spaceship, it can cause the leak of ions discussed in the earlier section on the point effect, thus reducing the performance of the motor or even stopping it if the flux is in the opposite direction of the movement.

You can, of course, make other types of spaceships as well. One suggestion is to adapt the system to a miniature of the *Star Wars'* pod racer, as shown in Figure 3.4.12, or of *Star Trek's Enterprise*.

But remember, the ionic motor is not powerful, and if the miniature pod racer is not light enough, it will move very slowly or not at all. Build your spacecraft to be as light as possible.

The support on which the spacecraft sits consists of a piece of rigid wire (16 or 14 AWG) 15 to 20 cm long. The wire is supported on a solid base. I used a small washer where the wire is soldered to the base, as shown in Figure 3.4.13.

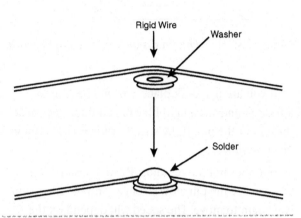

Figure 3.4.13 *The groove to support the spacecraft.*

A groove in the spacecraft is made to match the wire support. It is important to verify exactly the point where the groove in the spacecraft should be made. The spacecraft and the support must be perfectly aligned when the spacecraft is released from the support. You can use a small ball of solder on the opposite end of the wire as a balance weight.

Another important point to consider is dhat no metallic, pointed elements are present in this support. The only pointed element should be the needle in the ionic motor on the actual spacecraft. Figure 3.4.14 shows the final support for the spacecraft.

Testing

Place the spacecraft and its support in position. The craft must be free to move in the circular trajectory. Turn on the power. A soft noise will indicate that the circuit is in operation and producing high voltage. The neon lamp in the spaceship (if used) will glow.

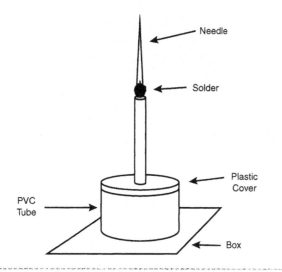

Figure 3.4.14 *The final support for the spacecraft.*

If you are in a dark place, you will be able to adjust P1 and see a light blue or red fluorescence at the end of the pin. If the spacecraft is very light, it will begin to move.

If it does not move, adjust P1 to a position where you can see the pin glow stronger and the spaceship will begin to move. The speed will depend on the weight of the ship and the power released by the high-voltage circuit. If your spacecraft still does not start, you must reduce the weight and/or the friction between the spacecraft and the support. Use a lighter or shorter wire for the support if necessary.

A Space Race

You can organize a competition with your friends to see who can build the fastest spacecraft powered by an ionic motor. Try limiting the length of the support wire to 20 centimeters for all competitors. The competitor whose spacecraft moves in its circular trajectory the most times in 1 minute wins the competition.

Exploring the Project

The high voltage produced by the electronic circuit can be used in many other experiments and applications. High-voltage experiments are very interesting when studying physics at the high school level. The following are some suggested projects.

Wireless Fluorescent Lamp

In this exercise, you can become an electronic magician by causing a fluorescent lamp to glow in your hands simply by using the high-voltage electric field produced by the circuit described here. It is recommended to work in a dark place, because the light produced by the lamp is typically not strong enough to be seen in brighter areas.

Take the lamp as shown in Figure 3.4.15 and place it near the high-voltage terminal. The lamp will glow.

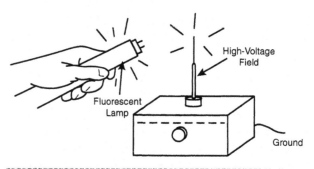

Figure 3.4.15 *A fluorescent lamp glows when placed near the electric field produced by the circuit.*

Caution! Do not touch the high-voltage terminal during the experiment. You will receive a shock. Please remember that although you are working with high voltages (in the range of thousands of volts), the discharge is not fatal. Still, the shock can be severe and hurt or burn if you touch any part of the circuit when in operation.

By passing your hands up and down along the sides of the lamp, the lamp will glow only in the parts your hand passes over. You can show that you are able to "control" the electricity in the lamp as if you were a real magician.

Jacob's Ladder

A classic experiment to show the effects of high-voltage sparks is called Jacob's Ladder. An electric arc begins at a small gap between the electrodes of a high-voltage source; it rises up along and between a pair of solid wires of gradually increasing separation (positioned with more space between the far ends than between the beginning ends). The electric field between the wires will decrease with this increasing separation.

The electric arc will blow out near the far end of the wires where the electric field is no longer sufficient to sustain the discharge, and the process will begin again. Figure 3.4.16 shows how the wires can be placed to perform this experiment.

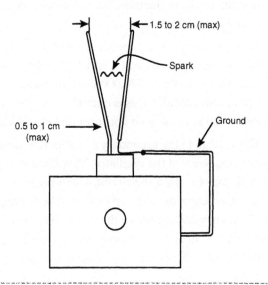

Figure 3.4.16 *Experiment with the separation between wires to obtain the best positioning.*

The amount of the wires' separation depends on the voltage produced by your circuit. Note that in normal conditions, the arc can reach 1 cm for each 10 kV of voltage.

Domestic Thunderbolts

Contrary to popular belief, common incandescent lamps do not have a vacuum inside. Instead they are filled with a noble gas such as nitrogen, argon, or neon. This avoids the danger of a critically low internal pressure producing an implosion, as can be caused when the bulb is knocked or falls.

This low-pressure gas becomes conductive with the addition of relatively low voltages. Therefore, by applying the high voltage generated by this circuit, the gas ionizes and a flux of ions carries the charges away from the lamp. Sparks leaving the support of the filament are seen in Figure 3.4.17. Touch the lamp with your finger and the sparks will concentrate in the place where your finger is. Because the glass is a good isolator for the charges, the sparks will stay inside the glass and you will not feel any shock.

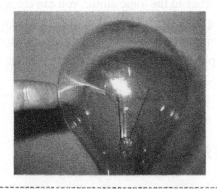

Figure 3.4.17 *Sparks jumping to the author's finger inside a common lamp. No shock is caused because the sparks actually stop inside the glass.*

Notice that very little current is required to flow across the filament to produce heat and light in the conventional way. The filament acts only as an electrode emitter of ions. The lamp, for best results, must be rated to 150 watts or more.

Electrostatic Experiments

Physics experiments with high voltages can be performed using the high-voltage generator of an ionic motor. Because many experiments need a continuous voltage and the ionic motor circuit generates an alternating voltage, a rectifier will need to be added. Figure 3.4.18 shows how the high-voltage rectifier can be added.

Any high-voltage silicon rectifier rated to 15 kV or more is suitable for this circuit. You can find a diode like this in old TVs and video monitors. Once you

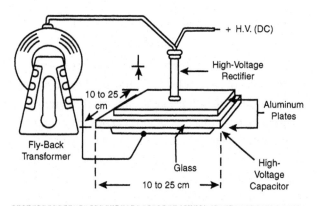

Figure 3.4.18 *Adding a high-voltage rectifier.*

add a rectifier to the ionic motor, you can charge a metallic sphere or experiment with electroscopes or other projects. (A simple electroscope built with metallic plates can be used to detect the charge of objects.)

Cross Themes

High-voltage experiments fit well with high-school-level curriculum for physics. The evil genius student (or his or her teacher) can use any of the projects in this section that have the following crossover themes:

- Showing the point effect
- Experimenting with gas ionization
- Performing experiments involving the action of sparks
- Generating continuous high voltage to charge objects and perform experiments in electrostatics
- Experimenting with calculation involving Coulomb's law

Additional Circuits and Ideas

High voltage can be generated from a power supply line in many ways. Circuits normally are high-power or low-power oscillators during power output stages. A variety of circuits can be used to generate high voltage. The following sections include suggestions of particular circuits that can be used in experiments with ionic motors.

High-Power Circuit

I built this circuit to show my students both the basic ionic motor and then the sparks generated in a lamp, as shown in Figure 3.4.19.

The basic high-voltage generator is an oscillator using a *complementary metal oxide semiconductor* (CMOS) *integrated circuit* (IC) 4093 and a power bipolar NPN transistor. The load for the oscillator is a coil in the ferrite core of a common high-voltage transformer (or flyback), such as the ones found in old TVs and computer video monitors. The same transformer used in the basic version can be used in this version too.

The low-voltage pulses produced by the oscillator induce in the secondary coil a very high voltage. Voltages between 10,000 and 30,000 volts are induced in common transformers. This is enough to create a flux of ions when applied in an appropriated electrode.

The voltage in the secondary coil depends greatly on the frequency of the oscillator. Therefore, it may be necessary to adjust the circuit. This adjustment can act as an accelerator to the spaceship. When changing the frequency and therefore the speed, the noise produced by the flux of ions is perfectly audible and goes from bass to treble (from low frequency to high frequency).

The power supply is formed by a transformer that reduces the AC power to about 15 volts × 3 amps. At the same time, the transformer acts as an isolating element between the circuit and power line, adding security to the project so no shocks occur if any part is touched. The circuit is safe.

The low AC voltage at the secondary coil of the transformer is rectified by diodes and filtered by a large electrolytic capacitor. Because the circuit is not critical, no voltage regulation is needed for the power stage. Only a small voltage regulator (7812) is used to power the IC.

The IC used as the oscillator is a 4093 (with four Schmitt NAND gates). One of the gates is configured as an oscillator with the frequency given by C3 and

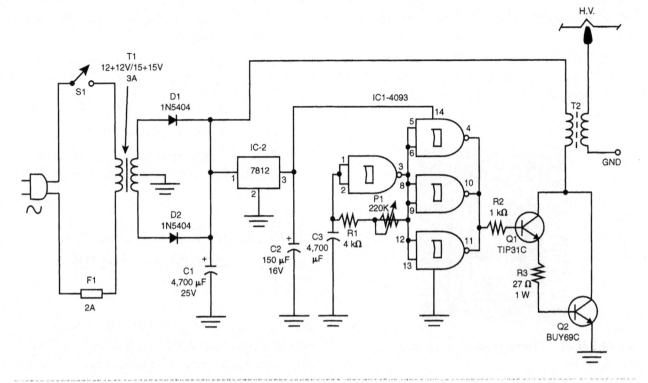

Figure 3.4.19 *Project with a high-voltage generator mounted inside a wooden box.*

the R network formed by P1 and R1. P1 is the acceleration control for the ionic motor.

The square wave produced by this circuit, ranging between 100 Hz and 10,000 Hz, is applied to the other three gates, which are configured as a digital, buffer amplifier. The power of the output of this stage is enough to drive the transistors.

The output stage is formed by two transistors: a driver (TIP31) and a high-voltage power transistor BUY69A. You can experiment to find appropriate substitutes for the high-voltage transistor. A transistor with gains above 20 and a Vce (max) over 200 volts might work well.

As you did earlier in the chapter when testing the circuit, you might want to alter C2 and C3 to find the best performance. C2 can be a capacitor in the range of 0.01 μF and 0.1 μF, and C3 can be a capacitor in the range of 0.100 μF and 0.560 μF.

I used a small universal PCB to fit the IC and then soldered a terminal strip at each side of the board. Other components, such as the diodes, power transistor Q1, and some resistors, were soldered to a terminal strip.

All the components fit into a wooden box. Figure 3.4.20 shows the parts placement diagram. Each power transistor is placed on a heatsink, and the transformer is kept firmly in place in the box with screws. The box used in the project measures 22 × 22 × 15 cm. This box was purchased in an office supply store and is typically used to hold file cards.

The flyback, usually taken from any TV or video monitor out of use, must be the type without the tripler circuit that uses diodes and capacitors. This kind of transformer is protected by a plastic cover to avoid any contact with the core. Therefore, it is

Figure 3.4.20 *Parts placement inside the box.*

impossible to add the primary coil as required in the project.

The primary coil is formed by 10 to 12 turns of common plastic-covered wire as shown in Figure 3.4.21. Operation tests can be made as described in the basic project.

Figure 3.4.21 *Primary coil of the flyback transformer.*

Parts List—High-Power Circuit

IC1	4093 CMOS IC
IC2	7812 12-volt voltage regulator IC
Q1	TIP31C NPN silicon power transistor
Q2	BUY69C NPN silicon power transistor
D1, D2	1N5404 silicon rectifier diodes
R1	4.7 kΩ × 1/8-watt resistor (yellow, violet, red)
R2	1 kΩ × 1/8-watt resistor (brown, black, red)
R3	27 Ω × 1-watt resistor (red, violet, black)
P1	220 kΩ linear or logarithmic potentiometer
C1	4,700 μF × 25-volt electrolytic capacitor
C2	4,700 *pico farads* (pF) ceramic or polyester film capacitor
T1	transformer, a primary coil rated according the power supply line; secondary coil rated to 12 + 12 volts × 3 amps CT
T2	flyback transformer (see text; same used in the original project)
S1	SPST switch (optional ganged to P1)
F1	2-amp fuse with holder

PCB, terminal strips, heatsink, screws, knob for P1, solder, power cord, wires, box, etc.

Relaxation Circuit Using a Silicon Controlled Rectifier (SCR)

Another circuit suitable for the ion IC motor is shown in Figure 3.4.22. Figure 3.4.23 shows the basic components soldered to a terminal strip. Some readers may want to mount the same circuit using a *printed circuit board* (PCB), if one is available.

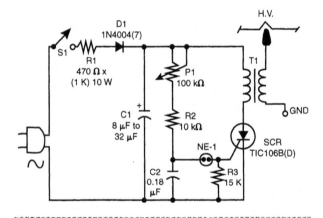

Figure 3.4.22 *Relaxation oscillator generating high voltage.*

This circuit is very simple but has two important limitations. First, it is powered directly from the power supply line, so the reader must take much care when using it. The second limitation is the frequency, which is limited to a few kilohertz due the characteristics of the *silicon controlled rectifier* (SCR).

This circuit operates as follows: The C1 capacitor is charged via the power supply line. At the same time, C2 charges via P1/R2 until the trigger voltage of the neon lamp is reached. At this moment, the lamp triggers the SCR on.

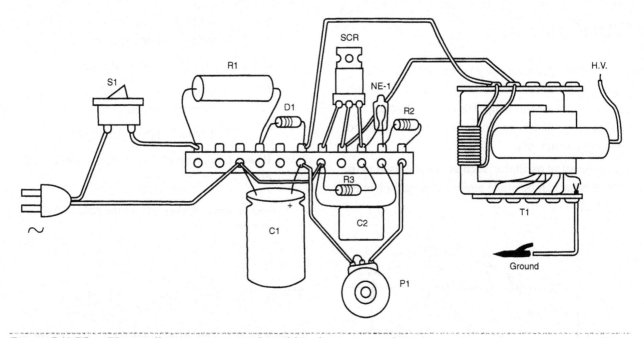

Figure 3.4.23 *The small components can be soldered to a terminal strip.*

Passing from the off state to the on state, the SCR allows the C1 capacitor to discharge through the primary coil of the flyback transformer. A high-voltage pulse is induced in the secondary coil.

As the capacitor discharges down to a level where the current of the SCR is holding steady, it turns off and a new cycle begins. The capacitors can also be charged again so that a new high-voltage pulse is produced. The repetition rate of the process is determined by adjusting P1.

L1, the primary coil of the transformer, is formed by 8 to 15 turns of 18 to 22 plastic-covered wire in the core of the flyback (as in the basic project earlier). C1 can have values in the range between 8 and 32 μF and a voltage rate over 200 volts.

The SCR doesn't need to be mounted on a heatsink. The pulses produced by the discharging capacitor are not long enough to generate a large amount of heat.

Notice that you can replace the flyback transformer with a car ignition coil in your experiments.

Parts List—Relaxation Circuit with SCR

SCR	TIC106B (D) SCR
D1	1N4004 (7) silicon rectifier diode
NE-1,	NE-2H or equivalent small neon lamp
R1	470 Ω × 10-watt (1 kΩ × 10-watt)* wire-wound resistor
R2	10 kΩ × 1/8-watt resistor (brown, black, orange)
R3	15 kΩ × 1/8-watt resistor (brown, green, orange)
C1	8 μF to 32 μF × 200-volt electrolytic capacitor
C2	0.18 μF × 100-volt ceramic or polyester film capacitor
P1	100 kΩ linear or logarithmic potentiometer
S1	SPST switch (optional)
T1	Flyback transformer

Terminal strip, power cord, wires, solder, knob for P1, wooden box, fuse holder, etc.

*These values are for the *220/240 volts of alternating current* (Vac) power line.

SIDAC Circuit

Silicon diodes for alternating current (SIDACs) are negative-resistance devices with characteristics similar to those of a neon lamp. The main difference is that they are solid-state devices that, when triggered, can conduct currents.

A relaxation oscillator, such as the one used in the SCR version, can be made easily, as we can see in Figure 3.4.24.

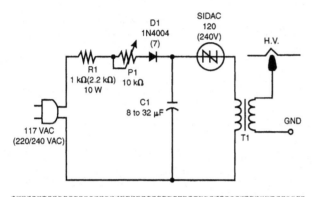

Figure 3.4.24 *High-voltage generator using a SIDAC.*

The circuit can be powered from 117 VAC or 220/240 VAC according to the used SIDAC. The operating principle is simple to understand: The C1 capacitor charges across R1/P1 until the trigger voltage of the SIDAC is reached. When triggered, the capacitor discharges across the primary coil of the transformer, producing a high-voltage pulse in the secondary coil.

The primary coil of the transformer is made in the same manner as in the project using the SCR: 8 to 12 turns of common plastic-covered wire. You can also replace the flyback transformer with a car ignition coil in this project as you might have done in the SCR project.

Parts List—SIDAC Circuit

SIDAC	NTE7419 (110 V), K1200 G (110 V), or K2400G (220 V)
D1	1N4004(7) silicon rectifier diode
R1	1 kΩ × 10-watt (2.2 kΩ × 10-watt)* wire-wounded resistor
P1	10 kΩ wire potentiometer
C1	8 to 32 µF × 200-volt (400-volt)* electrolytic capacitor
T1	Flyback transformer (see text)

Power cord, terminal strip, box, knob for P1, etc.

*Values between brackets are for the 220/240 VAC power line.

The Technology Today

Ionic motors are far from practical here on Earth or even as the only power source for crafts in outer space. The actual motors used in some applications in space are the low-power types; many are used only to change the position or trajectory of satellites in orbit. One example is the spacecraft *Deep Space 1*, released in 1999 by NASA. Figure 3.4.25 shows an ionic-motor-powered spaceship.

Figure 3.4.25 *An ionic spacecraft (Photo courtesy NASA).*

Even with the limitations, much research is being done to develop powerful motors or even motors that can stay in operation over very long periods of time. Using such motors, it will be possible to accelerate a spaceship during 1 year in a manner that will allow the spaceship to reach relativistic speeds far into the future, to the end of the time.

Those spaceships can be used to explore other solar systems, solar systems that are too far away to be reached by common rockets. NASA already has plans to build such spaceships.

Project 5—Experimental Galvanometer

Galvanometers are sensitive, electric current detectors. When the evil genius builds a galvanometer, he or she will also be able to identify some interesting mechatronic devices and perform experiments with alternative energy sources and photoelectric sensors.

The basic version of the galvanometer described here can detect currents as low as a few microamperes (mA), and with the aid of an electronic amplifier, even weaker currents can be detected. The high sensitivity achieved by using the additional circuit is perfect for projects like the following:

- Building a lie detector
- Identifying energy from experimental sources
- Building a light/dark detector

The project uses only a few components, most obtained from common sources such as old appliances and home objects, which will have no costs to the reader. Due the simplicity of the project, it is ideal for beginning evil geniuses, elementary students, and for teachers who want simple experiments to make their classes more interesting.

A cross theme found in this project was used by the classes at the school where the author teaches. Figure 3.5.1 shows the young students building their galvanometers to perform some experiments to detect electric currents. Experiments can be designed with different levels of difficulty according to the grade level of the students.

Objectives

Although it can be great fun detecting currents with the galvanometer, some important objectives should be considered when mounting this galvanometer. Readers can learn about the following themes:

- How to detect electric currents
- What Oesterd discovered about electric currents and magnetic fields
- How to build alternative sources of energy
- The operation principles of real galvanometers

The Project

The basic project is to build a sensitive galvanometer using only a small number of components. The project consists of creating a coil that involves a magnetic needle hung by a thin line. When an electric current passes across the coil, a magnetic field is created and its action moves the needle.

By the movement of the needle, the reader can evaluate the amount of current in the coil. It is possible to detect even very weak currents in the range of few microamperes or millionths of amperes. A current produced by two metallic plates inserted in a lemon is enough to move the needle of this sensitive detector.

How It Works

The Danish professor Hans Christian Oesterd discovered that electric currents can create magnetic fields.

Figure 3.5.1 *Students mounting the galvanometer as a cross theme (Photo courtesy Colegio Mater Amabilis).*

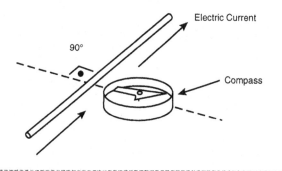

Figure 3.5.2 *Oesterd's experiment.*

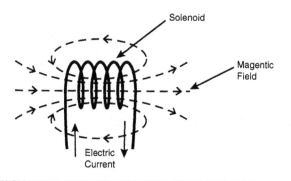

Figure 3.5.4 *The magnetic field is strongest inside a coil (solenoid).*

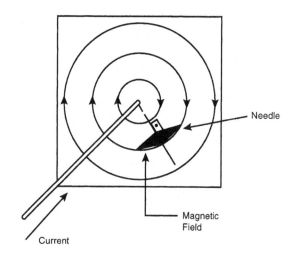

Figure 3.5.3 *The needle of the compass moved, stalling perpendicular to the wire.*

In his experiment, he placed a metallic wire near a compass, as shown in Figure 3.5.2.

When the switch was closed, the electric current flowing across the wire created a magnetic field strong enough to displace the needle of the compass. The needle moved, stopping in a position parallel to the force lines of the magnetic field (i.e., perpendicular to the field), as shown in Figure 3.5.3.

After this experiment, the researchers discovered that if the wire was wound into a coil, the magnetic field could be concentrated inside it, as shown in Figure 3.5.4.

As the reader can see, an electric current can move a needle to construct an electric current detector. This kind of instrument is called a *galvanometer*. Modern analog galvanometers consist of moving coils placed between the poles of a magnet, as shown in Figure 3.5.5.

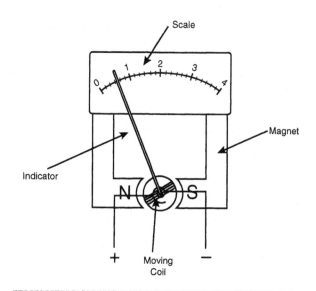

Figure 3.5.5 *A moving coil galvanometer.*

When the electric current flows across the coil, the magnetic field reacts with the field of the permanent magnet, creating a force that tends to rotate the coil. The force and the displacement are proportional to the amount of current and therefore to the displacement of a pointer along a scale.

Our galvanometer is very simple in its basic version but can be upgraded according to the skills and imagination of the reader. Although the sensitivity of our galvanometer is high, by adding electronic circuits, the sensitivity can be increased to an even greater level. Currents of a few microamperes can be detected by our more sensitive galvanometer.

Building the Galvanometer

Figure 3.5.6 shows the plans to build the galvanometer. The coil is wound around any cylindrical object with a diameter between 7 and 10 cm. You can use any wire you have on hand. You can choose from common plastic-covered wire (between 22 and 26 gauge) or enameled wire (between 22 and 28 AWG). Don't use very thin wires because they can't hold the needle.

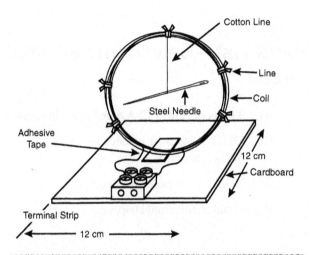

Figure 3.5.6 *The basic version of the galvanometer.*

The number of turns in your coil will determine the sensitivity and the electrical characteristics of your galvanometer. I recommend using 20 to 50 turns, depending on the wire you choose. To keep the turns coiled tightly together, you can use a string or an adhesive band once you have finished winding. The coil is then placed in a cardboard or wooden base. You can use glue or an adhesive band to secure it here.

Screws are used to connect the terminal strip that holds the coil to the external circuit. Remove the cover to connect the ends of the coil to the terminals. The enameled wires need to be uncovered as well, because the small layer of enamel acts as an isolator. Use a small pocketknife or wire cutters (and adult supervision if necessary) to remove the wire cover.

The next step in mounting the galvanometer is to prepare the needle. First, find a needle with a length between 5 and 7 centimeters. Rub it against a magnet to magnetize it. Then tie one end of a length of string to the needle. Tie the other end of the string to the coil.

Then hang the needle inside the coil in equilibrium. The needle should be at the center of the coil and without touching it. Figure 3.5.7 shows the galvanometer ready to be tested and used.

Figure 3.5.7 *The galvanometer ready to be used.*

Parts List—The Basic Galvanometer

5 to 15 meters of wire (see text)

1 needle, 5 to 7 cm long

1 piece of cardboard, 12 × 12 cm

1 terminal strip with screws

String, adhesive band, glue, etc.

Testing the Galvanometer

A simple circuit that allows you to determine the sensitivity of the galvanometer is shown in Figure 3.5.8.

First, adjust the potentiometer for the maximum value of resistance. This means you will need to move the cursor all the way to the right. Using a 10 kΩ potentiometer, the probe current should be about 0.3 mA or 300 μA.

Touching the terminal with the wire from the battery holder, as indicated in the figure, observe whether the needle moves. If it moves, your

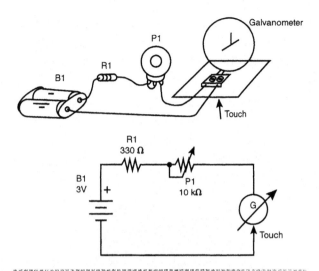

Figure 3.5.8 *The test circuit for the galvanometer.*

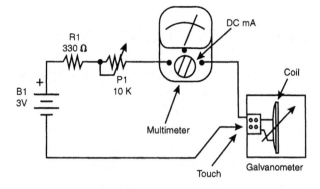

Figure 3.5.9 *Using a multimeter to determine the galvanometer's sensitivity.*

galvanometer is sensitive enough to detect a current as low as 300 μA.

If not, move the cursor of your potentiometer a few degrees to the left—about 2/3 of the complete rotation. By doing this, the resistance will fall to about 7,000 Ω. The probe current is now something less than 0.5 mA or 500 μA. Repeat the test.

If the needle moves, your galvanometer has a sensitivity of about 0.5 mA. If not, reduce the resistance of the potentiometer a little more, increasing the test current.

Place the cursor to 1/3 of the complete resistance, which places it in the range of 3,500 ohms. Considering the tolerances of the components, you will have a probe current of about 1 mA. Repeat the test, and observe if the needle moves. If the answer is yes, the sensitivity of your galvanometer is 1 mA. If not, increase the current again.

Now closing the potentiometer to have a zero resistance, the only resistance in the circuit is R1. With the 3-volt supply, the probe current will be 6 mA. Repeat the test. The needle must move now. If not, rebuild the unit, using more wire in the coil and verifying the line that sustains the needle.

For precision in testing your galvanometer, place a common multimeter in parallel and see exactly the current it detects. Figure 3.5.9 shows how this can be done.

Parts List—Galvanometer Test Circuit

B1 3 volts of power (2 AA cells and holder)

R1 330 Ω × 1/8-watt resistor (orange, orange, brown)

P1 10 kΩ linear potentiometer

Wires, solder, terminal strip, etc.

Exploring the Project

Experiments using the galvanometer can include some interesting projects. The experiments are divided in two groups: (1) using the galvanometer to detect the energy of alternative sources and (2) using the galvanometer as part of other projects.

Cross Themes

To present cross themes, some experiments use the galvanometer as detector of simple alternative sources of energy. Figure 3.5.10 shows two chemical alternative sources of energy.

Figure 3.5.10 (a) shows a cell made with a lemon or other citrus fruit. The electrodes are made of different materials such as copper for the positive, and zinc or aluminum for the negative pole. The voltage

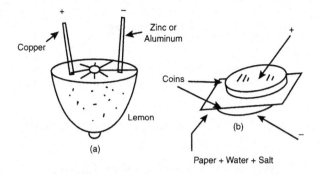

Figure 3.5.10 *Cells as an alternative source of energy.*

obtained from this cell is in the range from 0.3 to 1.0 volts depending on the metals used as electrodes.

Figure 3.5.10 (b) shows a cell made with two coins. They must be of different metals (copper, aluminum, silver, etc.). A silver coin with an aluminum coin can create about 1.0 volts. The paper between the coins is wet with a salt solution. Touching the terminals of these cells to the terminals of the galvanometer, the needle will move and show the voltage of the cell.

Another energy source that can be used to test your galvanometer is a photocell such as the ones found in hand calculators. Touching the terminals of the galvanometer with the wires coming from the power source (photocell) and exposing the cell to light, the needle will move.

Finally, you can use a small motor to generate electric energy from movement (see Project 8 to build an eolic generator). Figure 3.5.11 shows how the experiment must be done.

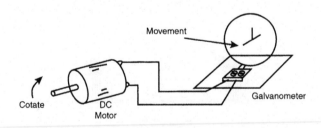

Figure 3.5.11 *Using a small DC motor as a dynamo.*

Increasing the Sensitivity

Although the original version of the galvanometer is sensitive enough to detect currents of a few microamperes, the reader can increase the sensitivity by adding an electronic amplifier.

The circuit uses one transistor and can be mounted on a terminal strip. Figure 3.5.12 shows the schematic diagram of the amplifier.

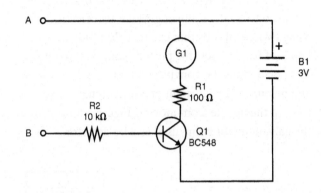

Figure 3.5.12 *Using an amplifier to increase the sensitivity of the galvanometer.*

This transistor allows the galvanometer to be used as a sensitive continuity tester and even as a lie detector. A current of only a few microamperes (1 to 10 depending on the sensitivity of your galvanometer) is enough to move the needle. Figure 3.5.13 shows how to mount the circuit using a terminal strip.

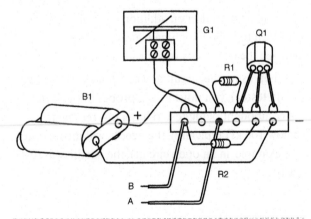

Figure 3.5.13 *The components are soldered to a terminal strip.*

You can test the sensitivity of this project by connecting a 1 MΩ potentiometer to the input. Test the circuit with the cursor of the potentiometer in different positions and determine the minimum resistance that causes the needle to move.

To use the galvanometer as a lie detector, you first need to know that the resistance of human skin changes when the person is under stress such as during an interrogation. The movement of the needle will detect small changes in the skin's resistance.

To create a lie detector from a galvanometer, you will need to do the following. Make two electrodes by using two metal rods that can be held firmly in the hands of the suspect (the person being interrogated). The person must be convinced that the circuit works, and he must keep constant pressure on the electrodes to keep the needle from moving. Figure 3.5.14 shows the galvanometer being used as a lie detector.

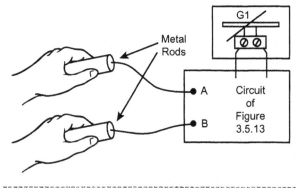

Figure 3.5.14 *Using the galvanometer as a lie detector.*

In the next section on additional circuit ideas, you will learn to build a high-gain amplifier that can increase the sensitivity of your lie detector.

When using the circuit as a component tester (continuity tester), you must attach probes to the input. If the component being tested has a low resistance (below 100 kΩ typically), the needle will move. If not (for example, in an open circuit), the needle will rest (stop).

Parts List—Amplifier

- Q1 BC548 or 2N2222 general-purpose *negative-positive-negative* (NPN) silicon transistor
- R1 100 Ω × 1/8-watt resistor (brown, black, brown)
- R2 10 kΩ × 1/8-watt resistor (brown, black, orange)
- B1 3 volts of power (2 AA cells and holder)
- G1 Galvanometer

Terminal strip, wires, solder, probes, etc.

Additional Circuits and Ideas

The basic galvanometer project can be changed, upgraded, or used for a different purpose. In this section, some variations on the galvanometer project are suggested.

Using a Compass

Figure 3.5.15 shows how a compass can be used to make a different version of the galvanometer. The coil uses enameled wire 30 to 32 AWG and is wound around a cardboard form with the dimensions shown

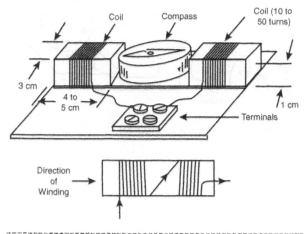

Figure 3.5.15 *Using a compass as a galvanometer.*

in the figure. Notice that the second coil must continue in the same direction as the first coil. If not, the magnetic fields created by the two coils will cancel each other, and the galvanometer will not function.

A High-Gain Amplifier

A high-gain amplifier using two transistors can increase the sensitivity of the galvanometer up to 10,000 times. The circuit for this amplifier is shown in Figure 3.5.16.

The transistors form a Darlington pair. The input resistance is very high, adding sensitivity to the circuit.

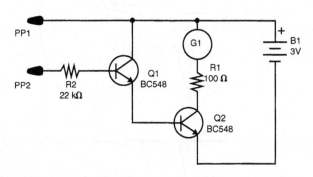

Figure 3.5.16 *High-gain amplifier.*

Parts List—High-Gain Amplifier

Q1, Q2	BC548 or 2N2222 general-purpose NPN silicon transistors
R1	100 Ω × 1/8-watt resistor (brown, black, brown)
R2	22 kΩ × 1/8-watt resistor (red, red, orange)
B1	3 volts of power (2 AA cells and holder)
PP1, PP2	Probes
G1	Galvanometer

Terminal strip or *printed circuit board* (PCB), wires, solder, etc.

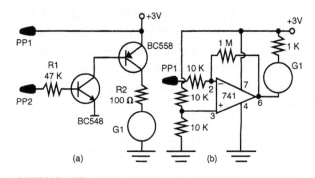

Figure 3.5.17 *Amplifier circuits for the galvanometer.*

Figure 3.5.17 shows some circuits that can be used to increase the sensitivity of the galvanometer.

In Figure 3.5.17 (a), a circuit is using a complementary pair of transistors (NPN/*positive-negative-positive* [PNP]). In Figure 3.5.17 (b), a circuit is using a Darlington transistor. In Figure 3.5.17 (c), the circuit uses an operational amplifier.

Light/Dark Sensor for Bio Energy

Light-dependent resistors (LDRs) are sensitive sensors for detecting light. They can be used to detect light and dark in experiments using the galvanometer as the indicator. A simple circuit used to detect light is shown in Figure 3.5.18.

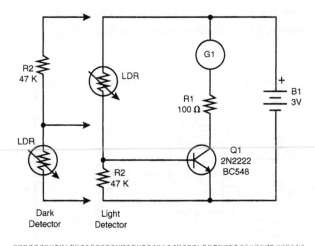

Figure 3.5.18 *Detecting light with the experimental galvanometer.*

Insert the batteries in the cell holder and put the needle of the galvanometer in a position perpendicular to the plane of the coil. Use a flashlight to illuminate the LDR. The needle will then move, indicating the light detection by the sensor.

Parts List—Light Detector

- Q1   BC548 or 2N2222 general-purpose NPN silicon transistor
- LCR Common LDR
- R1   100 Ω × 1/8-watt resistor (brown, black, brown)
- R2   47 kΩ × 1/8-watt resistor (yellow, violet, orange)
- B1   3 volts of power (2 AA cells and holder)
- G1   Galvanometer

Terminal strip, wires, solder, etc.

A Mobile Perpetual-Motion Machine

The circuit shown in Figure 3.5.19 will produce movements in the galvanometer's needle at regular intervals. The time between pulses can be adjusted by P1.

Because current drain is very low, the circuit can drive the galvanometer and keep it moving for months. This is really a quasi-perpetual-motion machine. You can attach the needle to some light-colored object, transforming it into a mobile as Figure 3.5.20 shows. Mobiles or perpetual-motion machines are devices that can operate without the need of energy. In the past, many persons believed it was possible to construct machines that needed no power to operate. Of course, this case is a simulation, since the conservation of energy would not allow it.

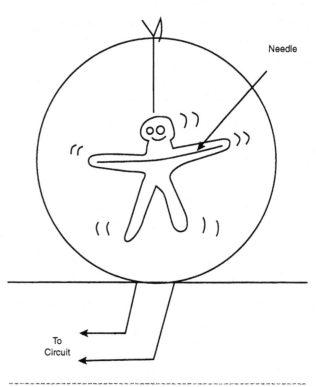

Figure 3.5.20 *A simple mobile.*

For testing, the circuit can be mounted on a solderless board or on a PCB. The best position for movement can be found by compensating for the magnetic field of the earth. If the needle tends toward very fast movements, increase the value of R4.

Parts List—Perpetual-Motion Machine

- IC-1   555 timer IC
- Q1   BC558 general-purpose PNP silicon transistor
- R1, R2 4.7 kΩ × 1/8-watt resistor (yellow, violet, red)

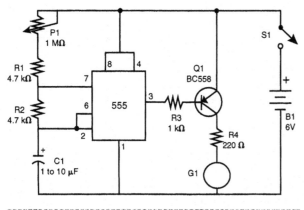

Figure 3.5.19 *Circuit of the perpetual motion machine.*

R3 1 kΩ × 1/8-watt resistor (brown, black, red)

R4 220 Ω × 1/8-watt resistor (red, red, brown)

C1 1 to 10 μF × 12-volt electrolytic capacitor

P1 1 MΩ trimmer potentiometer

B1 6 volts of power (4 AA cells and holder)

G1 Galvanometer

S1 SPST on/off switch

PCB, wires, solder, etc.

The Technology Today

Analog or mechanical galvanometers are not used in modern appliances. Today the old technology of moving coil indicators is found only in antique appliances, older cars, and industrial machines. More often, the detectors and indicators of today are digital. They don't use mechanical parts. Therefore, they are less likely to fail and can operate in any position.

Ideas to Explore

- Find galvanometers in old appliances and use them in the experiments described here.
- Design a mechanical system that moves a needle over a scale and then install it inside the coil.

Project 6—Experimenting with Electromagnets

Experiments with electromagnetism can be easily performed using simple homemade electromagnets. Small pieces of metal such as blades, nails, paper clips, screws, and other metallic objects will be attracted by the magnetic forces produced by an electromagnet.

The basic version of an electromagnet described here can be used in many experiments and practical applications. Only two components are used in this very inexpensive project. It is a project recommended to the evil geniuses who want to experiment with electromagnetism and need something very easy to build and use. Other versions, including a horseshoe electromagnet, will be described later.

Objectives

- Mount an experimental electromagnet.
- Learn how it works.
- Understand how some applications that use electromagnets work.
- Mount some mechatronic projects that use electromagnets.

The Project

The circuit works based on a discovery made in the nineteenth century by the Danish professor Hans Christian Oesterd. Oesterd discovered that when an electric current flows through a wire, a magnetic field appears around the wire. More details about Oester's discovery can be found in the previous project on the experimental galvanometer.

In few lines we can say that if a large amount of current flows through the wire, the field will be strong enough to move the needle of a compass, placing it perpendicular to the wire, as shown in Figure 3.6.1.

Section Three   The Projects

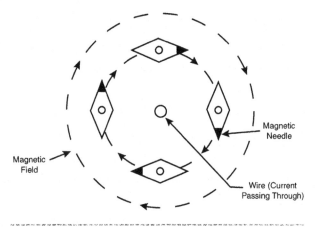

Figure 3.6.1 *The needle aligns with the lines of the magnetic field.*

The next step in his discovery was to demonstrate that the magnetic field could be concentrated if the wire is used to form a coil. Flowing through a coil an electric current can produce a strong magnetic field inside it. Figure 3.6.2 shows that the field is stronger inside the coil.

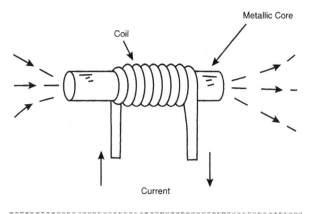

Figure 3.6.2 *Magnetic field inside a solenoid crossed by an electric current.*

Metal pieces placed at the center of the coil will remain magnetized, attracting small pieces of metal. But the interesting fact to be observed in this experiment is that the magnetic field exists only during the time in which the current is flowing. As long as the current is cut, the magnetic field disappears and no more attraction is observed.

Our simple project consists of a small electromagnet powered from a D cell. The field is strong enough to attract small metal pieces in demonstrations or even some interesting mechatronic applications.

Building the Electromagnet

The circuit is very simple: Only two components are needed. Figure 3.6.3 shows the schematic diagram of the electromagnet. Figure 3.6.4 shows the components used in this project and how they are wired.

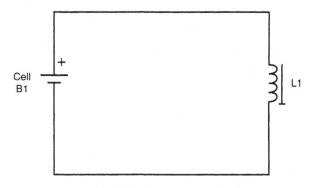

Figure 3.6.3 *Schematic diagram for the electromagnet.*

Figure 3.6.4 *The electromagnet wound to a nail and powered by a D cell.*

The electromagnet is made by winding 100 to 500 turns of any 28- to 32-gauge enameled AWG wire around a small nail or screw. Any iron nail or screw from 2 to 3 inches long can be used.

You must uncover the wire using a wire cutter, pocketknife, or other tool. The enamel is an isolator that will not allow the electricity to flow from the wire to the poles of the cell.

The power supply is a D cell; the electromagnet circuit drains a large amount of current, and AA cells will run down quickly if used.

Test and Use

Using the electromagnet is very simple: Place the ends of the wire in contact with the cell's poles. Verify that the cover of the wire has been removed at the point where it will contact with the cell so that the current flows. Remember that putting the electromagnet near small pieces of metal will attract them.

Don't use the electromagnet for extended periods of time. If the current flows through the coil for periods of time longer than a few seconds, the current will produce heat in the coil and damage it. Turn the electromagnet on and off in short spurts, only for the demonstrations.

Parts List—Electromagnet

- B1   D cell (see text)
- L1   Electromagnet (see text)

Exploring the Project: Other Versions

Figure 3.6.5 shows how to improve the performance of the electromagnet or to alter it using other formats and materials.

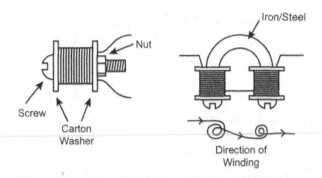

Figure 3.6.5   *Other versions of the electromagnet.*

Powering from a Different Power Supply

If your electromagnet is powered by a 3- to 6-volt power supply, add a 22-ohm × 1-watt resistor in series with the electromagnet. This resistor will limit the current flow through the coil, avoiding overheating.

Building a Crane

A mechatronic project using the electromagnet is shown in Figure 3.6.6. The PWM control can be used to control the speed of the up and down movement of the electromagnet. The S1 switch is used to catch or release objects.

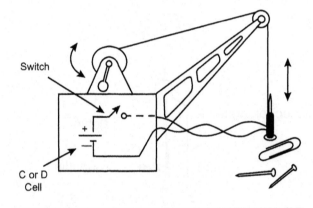

Figure 3.6.6   *Mechatronic crane using the electromagnet to move small metallic pieces.*

The Bottle's Challenge

Use this as a fun intelligence test you can give to your friends or parents. Put an empty bottle on a table with a metallic paper clip or a piece of metal (a nut or a screw, for instance) inside the bottle. Spread across the table small objects, such as a pencil, a piece of paper, a spoon, drinking straws, a small nail, 3 meters of enameled wire (28 to 32 AWG), coins, paperclips, cotton swabs, wooden matches, a D cell, and so on.

The challenge is to extract the object inside the bottle without moving or touching the bottle. The bottle can't be turned upside down and it can't fall from the table.

Of course, one of the first ideas that occurs to the person is to build some kind of hook using paperclips, attaching them to the wire, and trying to catch or hook the object in the bottle.

After many unsuccessful tries, the individual will be surprised when you simply wind the wire around the nail, grasp the ends of the wire to create a good electric connection, and connect it to the cell, making a magnet. Inserting the magnet into the bottle, the object will be attracted and be pulled easily to the opening, solving the problem. Figure 3.6.7 illustrates this being done.

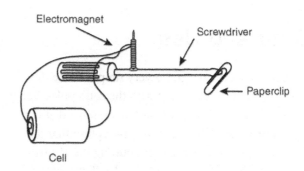

Figure 3.6.8 *A screwdriver is used to show the magnetization by induction.*

Additional Circuits and Ideas

Some additional circuits can be designed starting from the basic project. Some ideas are given in the next section.

Controlling the Power

You can control the amount of power in your magnet by using two circuits described in this book. The first uses an electronic potentiometer (see Project 7) wired in series with the electromagnet. Power the circuit with a 3- to 6-volt power supply and then add a 10-ohms × 1-watt resistor to limit the current, as shown in Figure 3.6.9.

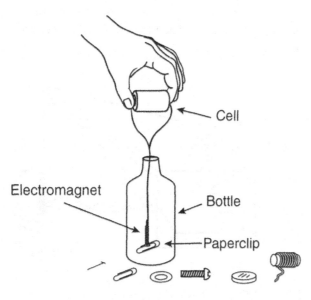

Figure 3.6.7 *Solving the problem of the bottle.*

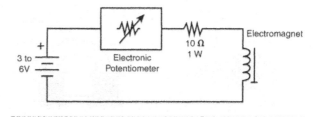

Figure 3.6.9 *Using a linear control.*

Cross Themes

Show that magnets attract only certain kinds of metals. They can't attract metals such as copper or aluminum. Explain why paper, glass, plastic, and other materials are not attracted by the magnets. Figure 3.6.8 shows a simple experiment to demonstrate the process of magnetization by induction. When the electromagnet is powered on, you will be able to attract a paperclip using the screwdriver.

The second project that can be used to control the power applied to the magnet is the *pulse width modulation* (PWM) control (see Project 3). In this case, you also have to limit dhe current by using a resistor (10 ohms × 1 watt). Figure 3.6.10 shows another circuit where the power source is being controlled. It has a constant current source using an LM350T IC.

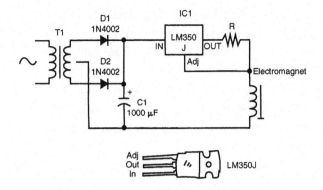

Figure 3.6.10 *Constant current source for experiments using magnets.*

Typically, R can be calculated for source currents in the range of 50 to 500 mA to a magnet according to its size and the power needed. To determine the current, you only have to divide 1.25 (the internal reference of the IC) by the current you need.

For instance, if you want to power the magnet described here with 100 mA (a recommended current for the project), R must be as follows:

$R = 1.25/0.1\ (100\ mA = 0.1\ A)$

$R = 12.5\ \Omega$

Use a 12.5- × 2-watt resistor. You can use the LM350T to source currents of up to 3 A, but with more than 200 mA the device must be placed on a heatsink.

Figure 3.6.11 shows how to mount a constant current source using a terminal strip as the chassis.

Parts List—Constant Current Source

IC1	LM350T
D1, D2	1N4002 silicon rectifier diodes
T1	Transformer: primary coil rated to 117 VAC; secondary coil rated to 7.5 + 7.5 volts or 9 + 9 volts × 1 amp
C1	1,000 µF × 16-volt electrolytic capacitor
R	Resistor, according the load current (see text)

Power cord, terminal strips, solder, wires, etc.

Changing the Poles

You can create an interesting experiment to show how the poles of an electromagnet can change by inverting the direction of the current in its coil. The experiment is shown in Figure 3.6.12.

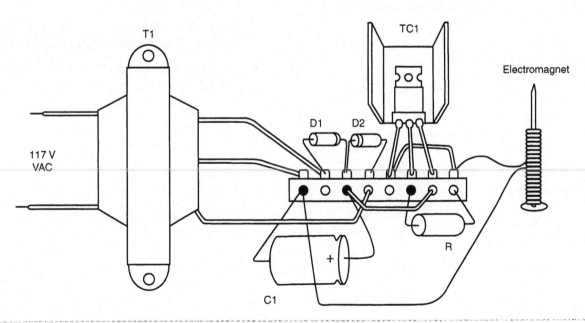

Figure 3.6.11 *Using a terminal strip to hold the small components of the project.*

Section Three   The Projects

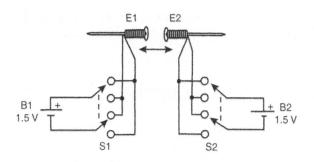

Figure 3.6.12 *Change the poles to show that poles of the same name repel one another and poles of opposite names attract one another.*

If the poles are opposites, the electromagnets will attract one another. If the direction of the current is changed in one of the poles, the poles will be the same and the electromagnets will repel one another.

The S1 and S2 switches are used to change the direction of the current, and the resistors are used to limit the current. Without the switches and resistors, the batteries will run down quickly, and the magnet will overheat.

Another Power Supply

Figure 3.6.13 shows how you can build a small power supply to power your experimental magnets and save your cells. The resistor is used to limit the current across experimental magnets, which tend to have low resistance and then drain large amounts of current.

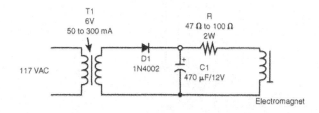

Figure 3.6.13 *Simple power supply to power experimental magnets.*

Using this circuit, your experiments will be safe and you can save your cells. The power supply in this circuit is not regulated, and its voltage can change according the load. Therefore, this circuit is not suitable for electronic applications. You can, however, use it to power small motors, solenoids, and other appliances that are tolerant to a larger range of voltages.

The Technology Today

Electromagnets are important elements of many modern devices. Motors, relays, and solenoids are examples of devices that use electromagnets.

Even in the industrial sector, powerful electromagnets are used to move pieces of metal with weights of up to several tons. They are very efficient for this task because they don't need a place to attach to the object being moved. They require only a place for contact with the object, that the current be turned on, and that is all. The object will be firmly attached to the electromagnet and can be moved.

Ideas to Explore

- Imagine the design of a simple telegraph that uses electromagnets.
- Look for information about the experiments made by Ampere with electromagnetism.
- Organize a competition to see who can build the most powerful electromagnet.

Project 7—Electronic Potentiometer

As we have seen in Project 3 on *pulse width modulation* (PWM) motor control, two types of motor controls exist: analog and digital. The analog control covers the amount of voltage applied across the motor, and the digital control deals with the characteristics of pulses applied to the motor.

PWM is a type of digital control and offers many advantages over other types of controls. But if you want something less advanced to control the speed of small DC motors, the light of an incandescent lamp, the force of a solenoid, or the temperature of a heating element, you can use an analog control.

In this project we are going to describe an analog power control or simply an electronic potentiometer using only three components. You will be able to experiment and build some mechatronic appliances based on the project with this type of power control. It's really simple. You need only attach the control between the fixed power supply, such as a battery, and the load to be controlled to get a perfect control of the amount of current flowing across the circuit.

Objectives

- Have a complete control for DC loads like motors, lamps, *light-emitting diodes* (LEDs), and so on.
- Do experiments with loads that need a power control.
- Build mechatronic projects using DC loads that need a power control (e.g., elevators, race cars, and fans).

How It Works

The circuit in this project uses a common, medium-power, *negative-positive-negative* (NPN) transistor to control the current flowing across a load. The control element is a carbon potentiometer.

Before we go any further, let's look at the differences between a potentiometer and a rheostat. A rheostat is a variable resistor connected so that it controls the amount of current in the circuit, and some types of potentiometers are specifically designed to be used like a rheostat. These devices have only two terminals. See Project 3 on PWM control of DC motors for more details about the operation of this circuit.

Some applications require a variable resistor, which has three terminals. When we wire components using all three terminals, we say that they are working as potentiometers. A potentiometer measures and divides the voltage. But we can also use a three-terminal variable resistor as a rheostat by wiring only two terminals, as shown in Figure 3.7.1.

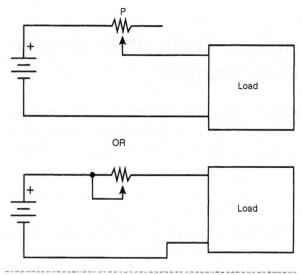

Figure 3.7.1 *Using a potentiometer as a rheostat.*

This project is used to control the amount of current in a circuit and requires only two terminals; therefore, we can call the project a rheostat. However, as we are using an electronic component called a potentiometer as base of the design, we can also call the project an electronic potentiometer. The reader will not be wrong to use either of the names.

The current flowing through the potentiometer is very low. It follows that the power dissipation is also

Project 7 – Electronic Potentiometer

very low and that no special type of potentiometer, such as a wire-wound potentiometer, is required.

In this kind of project, the amount of current that can be controlled depends primarily on the transistor used. The project given here uses a BD135 transistor, which can control up to 500 mA. However, the reader has some options to use a more or less powerful transistor, as shown in the following table. The circuit will operate using power sources of up to 12 volts.

Transistor	Maximum Load Current	R1/P1
2N2222, BC548	100 mA	220 Ω 1/2 W/4.7 kΩ
BD135, BD137, BD139	500 mA	100 Ω 1W/ 1 kΩ
TIP31/A/B/C	2 A	100 Ω 2W/1 kΩ*
2N3055	5 A	100 Ω 5W*/ 1 kΩ*

*Wire wound

How to Build an Electronic Potentiometer

Figure 3.7.2 shows the schematic diagram of the basic version of an electronic potentiometer using a BD135 *negative-positive-negative* (NPN) silicon transistor.

As only a very few components are used, they can be soldered to a terminal strip, as shown in Figure 3.7.3.

The transistor should be mounted on a heatsink. A piece of metal board screwed on the transistor is a good way to create a heatsink for this project. However, if you intend to control heavy-duty loads (above

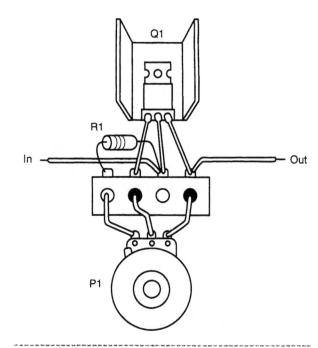

Figure 3.7.3 *The simplest way to mount the components is by soldering them to a terminal strip.*

1 amp) using powerful transistors, it is a good idea to mount the transistor in a large commercial heatsink.

It is important to verify the terminal's placement if you are using a transistor other than the one recommended in the basic project. Each transistor can have its own pin placement. An example is shown in Figure 3.7.4.

The electronic potentiometer can be housed in a small plastic box or added to the box housing the circuit that is to be controlled. Terminal strips or alligator clips are suggested to connect the electronic potentiometer to the power supply and to the circuit to be controlled.

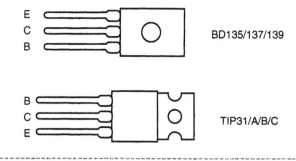

Figure 3.7.4 *Terminal placement for equivalent transistors.*

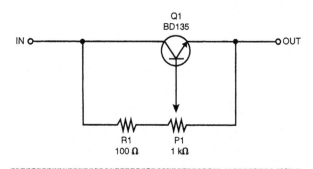

Figure 3.7.2 *Schematic diagram for the electronic potentiometer using an NPN transistor.*

68   Mechatronics for the Evil Genius

Using the Potentiometer

The circuit will operate using input voltages between 3 and 12 volts and current loads up to 1 amp. Wire the inputs IN+ and IN- to the power supply. Take care to position them correctly as those wires are polarized. Terminals OUT+ and OUT- should be wired to the load to be controlled.

Monitor P1 to see how the amount of power applied to the load changes. You can use a small lamp (12 V × 200 mA) to test the potentiometer. The changes in the applied power will cause the lamp to change its brightness.

If the transistor heats up too much when controlling a load and the load does not reach the maximum power, don't use the potentiometer with it. The current drained by the load is greater than the maximum current the potentiometer can supply.

Parts List—Electronic Potentiometer

- Q1 BD135 medium-power NPN silicon transistor (see text for equivalents)
- R1 100-ohm, 1-watt, 5 percent resistor (brown, black, brown)
- P1 1,000-ohm potentiometer, carbon type

Terminal strip, heatsink for the transistor, knob for the potentiometer, plastic box, wires, solder, etc.

Exploring the Project

This project is not built to be used alone. It is designed to be inserted between a power supply and a load to be controlled. Figure 3.7.5 shows a typical use of the electronic potentiometer.

The reader can now use his or her imagination and skills to design a good project to be controlled by this potentiometer and create more interesting experiments to learn more about mechatronics and science. We will give some suggestions.

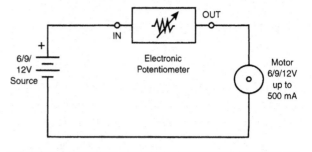

Figure 3.7.5 *Using the electronic potentiometer.*

Light Source for the Microscope

Many low-cost microscopes use as their light source an ambient light focused by a small mirror or a lamp. These are often powered by one or two cells. The light from these sources is not controlled and can cause some inconvenience to the user.

A good project for the evil genius who has a microscope is to build a controlled light source using the electronic potentiometer as a dimmer. The complete circuit for a 6 V × 200 mA lamp is shown in Figure 3.7.6.

The energy source can be four AA cells or four D cells, or the circuit using a transformer can be plugged into the AC power line. This circuit can also be used to perform experiments in optics if the reader is an amateur scientist or a teacher looking for cheap resources for your lab.

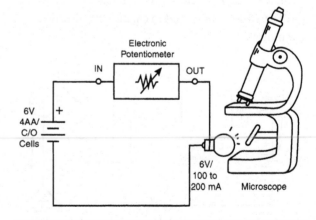

Figure 3.7.6 *Using the electronic potentiometer as a light dimmer for a microscope.*

Controlling a Small Heater

As a crossover into biology, your electronic potentiometer can be used to create a temperature control or ambient light control required for growing a culture or supporting the habitat of another living being.

A 22 Ω × 10-watt wire-wound resistor can be used as a heater (see Figure 3.7.7). Place it in the box where the temperature must be raised, and the heat can be controlled with the potentiometer. You can replace the resistor with a lamp and perform experiments with plant growth in a controlled environment.

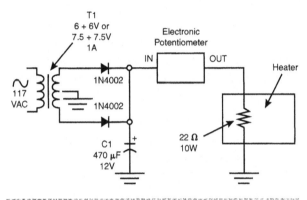

Figure 3.7.7 *Controlling a heater.*

Simple Elevator

Figure 3.7.8 shows how to use the electronic potentiometer to control the speed of a mechatronic elevator. The switches are used to change the direction of the motors.

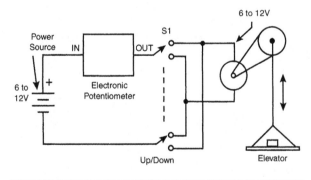

Figure 3.7.8 *Using the potentiometer in an elevator.*

Use a gearbox or other mechanical system to reduce the speed and increase the torque of the motor.

Cross Themes

Calculating Dissipation

As a crossover theme into physics, the electronic potentiometer can be used to teach how Joule's law is applied to a voltage divider. Each evil genius will have to calculate the amount of power dissipated by the transistor and the amount of power delivered to the load under a specified voltage. A simple multimeter can be used to perform this scientific experiment, as shown in Figure 3.7.9.

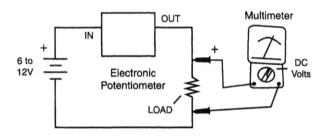

Figure 3.7.9 *A multimeter is used to measure the voltage applied to the load.*

The electronic potentiometer can also be included in experiments as crossover in the following subject areas:

- Joule's law
- Voltage dividers
- Association of resistors in series
- Measuring the torque of a motor

Additional Circuits and Ideas

The basic configuration for an electronic potentiometer can be changed and upgraded in several ways. For example, a more powerful transistor, Darlington circuits, or *integrated circuits* (ICs), can be used. Other suggestions are offered in the following paragraphs.

Using a PNP Transistor

The only difference between a circuit using an NPN transistor and a circuit using a *positive-negative-positive* (PNP) transistor is the direction of the current flow. Figure 3.7.10 shows how you can build an electronic potentiometer using a PNP transistor.

Using a Darlington Transistor

The use of a Darlington transistor allows higher values for the potentiometer. This means less heat is generated, and therefore a low-dissipation (carbon) type of potentiometer can be used. Recommended types for this application include the TIP115/116/117, which can control currents up to 1.25 amps, and the series TIP120/121/122, which can control currents up to 3 amps. Remember to always test and practice with less current. Figure 3.7.11 shows the circuit.

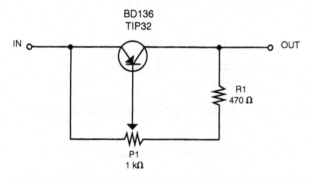

Figure 3.7.10   *Using a PNP bipolar transistor.*

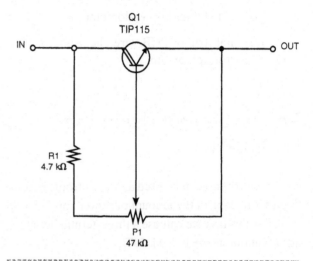

Figure 3.7.11   *Electronic potentiometer using a Darlington transistor.*

For currents up to 500 mA, you can use the BD136/138 or BD140. For currents up to 2 amps, you can use the TIP32/A/B or C. In any case, the transistor must be mounted on a heatsink.

Another option for incorporating a Darlington configuration is to use two transistors connected as a Darlington pair (see Figure 3.7.12).

Parts List—Electronic Potentiometer Using a PNP Transistor

Q1   BD136 medium-power NPN silicon transistor

R1   470 Ω × 1/2-watt resistor (yellow, violet, brown)

P1   1 kΩ linear or logarithmic potentiometer

Heatsink for the transistor, wires, solder, *printed circuit board* (PCB) or terminal strip, etc.

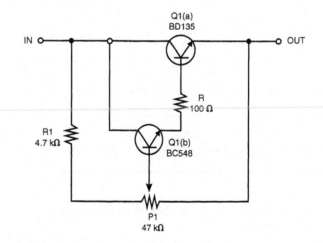

Figure 3.7.12   *Two NPN transistors can be coupled to form a Darlington pair.*

Again, when using the Darlington pair configuration, the potentiometer can be a carbon type, because the amount of current flowing across it is very low.

Parts List—Using a Darlington Transistor

- Q1 TIP115 or BC548/BD135 Darlington transistor or Darlington pair (see text)
- R1 4.7 kΩ × 1/8-watt resistor (yellow, violet, red)
- P1 47 kΩ linear potentiometer

Terminal strip or PCB, wires, heatsink for the transistor, solder, etc.

Power Control Using the LM350T IC

A voltage control can be used as an electronic potentiometer to control the amount of power applied to a load. For this task we can use a three-terminal voltage regulator as the IC LM350T.

This IC can source up to 3 amps to loads, applying voltages in the range between 1.2 and 32 volts. Figure 3.7.13 shows how to wire this IC in a voltage control.

The main difference when using this circuit is that the adjust terminal must be connected to the ground. This means that, instead of two terminals, as in the original version, this circuit has three terminals.

When using this circuit, you must plan on a drop in voltage of about 2 volts in the IC. This means that when you require 12 volts for a load, the input voltage must be at least 14 volts.

Finally, we must observe that the minimum voltage applied to the load is not 0 volts but 1.2 volts. This is due to the internal reference of the IC, which has a zener diode of 1.25 volts. If the application needs 0 volts as its minimum, this circuit is not suitable.

Parts List—Power Control Using the LM350T

- IC-1 LM350T IC
- R1 220 Ω × 1/2 W resistors (red, red, brown)
- P1 4.7 kΩ linear or logarithmic potentiometer

Heatsink for the IC, wires, solder, etc.

A Constant Current Source

An important method used to control current loads is the constant current source. By using a DC motor, it possible to keep the current constant, thus compensating for changes in the load.

Therefore, with a constant current source, the power applied to the motor doesn't have to change due to a change in the load when the motor slows down. The circuit for this application is shown in Figure 3.7.14.

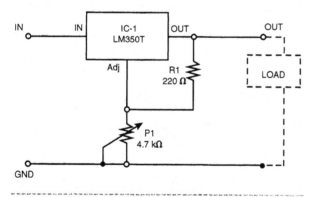

Figure 3.7.13 *The LM350T as an electronic potentiometer.*

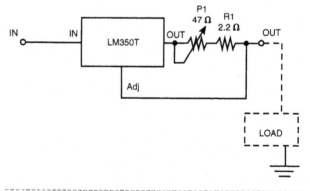

Figure 3.7.14 *A constant current source used as an electronic potentiometer.*

R1 is calculated according the maximum current in the load. The following formula can be used for this calculation:

$$R1 = 1.25/I$$

where I is the maximum current in the load.

It is important to observe that the minimum current in the load is not zero but is determined by the internal value of the potentiometer. For instance, when using a 47 Ω potentiometer, the minimum current will be:

$$I = 1.25/47 = 0.026 \text{ A or } 26 \text{ mA}$$

In other parts of the application, this current is not enough to move the motor or to overheat it.

The input voltage V_i must be 2 volts more than the nominal voltage of the load. For instance, for a 6-volt DC motor, V_i must be 8 volts. The IC LM350T must be mounted on a heatsink, and the maximum load current is 3 amps.

Parts List—Constant Current Potentiometer

IC-1 LM350T IC voltage regulator

R1 2.2 Ω × 1-watt wire-wound resistor for 500 mA motors (see text for other currents)

P1 47 Ω wire-wound potentiometer (see text)

Terminal strip or PCB, heatsink for the IC, wires, solder, knob for the potentiometer, etc.

The Technology Today

Linear power controls like the electronic potentiometer in this project are not common today. Controls for toys and for DC lamps in car dashboards are examples of products that use this type of control. PWM controls, such as the one shown in Project 3, and AC power controls, such as the ones used with lamps and motors, are much more efficient.

These PWM and AC controls are based on oscillator circuits or power semiconductors, such as those in special SCRs and TRIACs.

Project 8—Experiments with Eolic Generators

Alternative sources of energy are the target of much research all over the world. With the natural energy resources diminishing, some of the focus has turned to finding new sources of energy that can satisfy our demand and provide energy quickly and inexpensively.

One of the most important alternative energy sources is the wind. Fast-moving air (wind) represents an available pressure gradient great enough to power conversion-of-energy devices.

The name of the alternative source of energy represented by the wind is *eolic* from the Greek word *Eos*, meaning wind. Eolic generators are devices or transducers that convert eolic energy, or the energy of the wind, into electricity.

This project describes a simple eolic generator with an output great enough to power small electronic devices. This section also describes variations on the project for increasing the generator's power and for building other devices that can be used in heavy applications as alternative energy sources.

Our project will allow readers to save costs related to the use of traditional electrical energy sourced through the local energy distributor and to use some of the free power delivered by Mother Nature.

Section Three   The Projects

Our basic eolic project is very simple and will produce only enough energy to show you that it is possible to make the conversion. A variety of experimental circuits are suggested that can be powered by your eolic project. You will learn a good deal about eolic energy and even create a lighting system, a simple radio, a *light-emitting diode* (LED), and other simple devices that can be powered by your alternative energy source.

Objectives

- Show how the power of the wind can be converted into electrical energy.
- Use a small DC motor as a dynamo to power small electronic appliances.
- Learn how a dynamo works.
- Have an idea about the amount of wind power that is possible to convert into electrical energy.
- Know how to design a powerful eolic generator.

The Project

Figure 3.8.1 shows that a small eolic generator can be built using any DC motor.

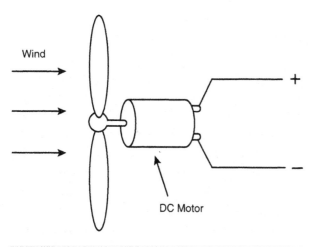

Figure 3.8.1 *Coupling a fan to a DC motor, it is converted into an eolic generator.*

Applying enough wind to move to the fan, the power released by the wind via the fan is converted into electrical energy. The amount of energy delivered by the system depends on several factors such as the following:

- The size of the fan (or the amount of wind that can be collected)
- The size, or the power, of the motor used as generator
- The amount of wind or speed of the wind in the place where the eolic generator will operate
- The efficiency of the system when transferring the fan movement to the motor

By combining these four factors, it is possible to predict the ideal size of the fan for a particular DC motor (generator), to know the amount of energy that will be released, and to know what is ideal for the application the reader has in mind.

Small DC motors don't require more than a fraction of a milliwatt of electrical power. However, this small amount is enough to power a small experimental radio, light an LED or small lamp, or charge a battery.

The ability to charge a battery can be especially important: You can charge the battery of a lantern, cellular phone, radio, or other appliance in order to use them when you need them. If you find yourself in a place where other energy sources are not available, the solution of an eolic generator should be considered.

Important information is given later in this project for the evil genius who wants more than just the few milliwatts of power generated by the basic project.

How It Works

As you learned in the previous projects on galvanometers and electromagnets, when a current passes across a wire, a magnetic field is created. This phenomenon can be used in several devices such as motors, electromagnets, solenoids, and relays.

In a DC motor, for instance, the magnetic field created by a current can be transformed into real movement when it interacts with the field of magnets (or the field created by other electromagnets).

The important observation regarding magnetic fields that are produced by an electric current is that an inverse effect is noted: When a magnetic field acts on a wire, a current is induced. Figure 3.8.2 shows that if a magnet moves near a wire, a current is induced, flowing across an external circuit.

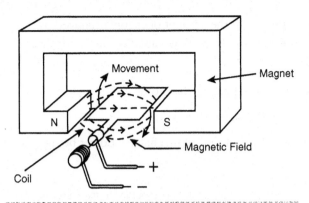

Figure 3.8.3 *The structure of a dynamo.*

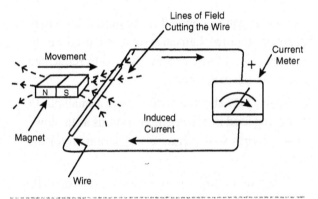

Figure 3.8.2 *A current is induced by a variable magnetic field.*

Observe that the lines of the magnetic field must cut the wire to cause the induction. If the magnet moves parallel to the wire, no induction occurs.

However, the featured fact in this phenomenon is that the power released to move the magnet is converted to electrical power, released by a current across the external circuit.

To increase the amount of power induced in this process, a coil can be used instead of a single wire. When the magnet passes near the coil, many turns are cut at the same time and therefore the induction process is magnified, creating a greater amount of electrical power.

This is exactly the principle behind the operation of a dynamo or alternator. A coil turns inside a magnetic field created by magnets. When crossing the magnetic lines of the field, electrical energy is generated. Figure 3.8.3 shows the structure of a dynamo.

An important fact to keep in mind about dynamos is that they don't create energy; they only transform mechanical energy into electrical energy. This means that the amount of electrical energy produced by a dynamo is always less than the amount of mechanical energy supplied to it, because no generator can convert 100 percent of the power.

A DC Motor as a Dynamo

In a small DC motor, the electrical current flowing across the coils creates a magnetic field that interacts with the field of a permanent magnet, producing the power that moves the rotor.

This architecture can operate in the inverse mode as well. When forcing the rotor to turn, the coils will move, cutting the magnetic lines of the magnetic field created by the magnets. This means that electrical energy is induced in the coils. This energy can be conducted to an external circuit, as shown in Figure 3.8.4.

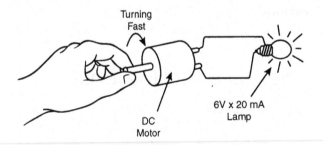

Figure 3.8.4 *When turning a DC motor, it can transfer electrical energy to an external circuit.*

Small DC motors rated at voltages from 3 to 12 volts can conduct currents up to 200 mA according to their size.

Of course the amount of energy produced by a motor, designed as a dynamo, depends on several factors:

- The speed of the motor
- The size of the motor
- The amount of mechanical power being sourced

The idea behind this project is to use wind as an alternative energy source, but the reader should also consider exploring energy from other sources. This could entail water flowing (a creek or waterfall) or a mechanical system (mechatronic) controlled by your own muscular power or some animal, as is suggested by Figure 3.8.5.

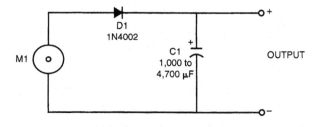

Figure 3.8.6 *Basic circuit for the eolic generator.*

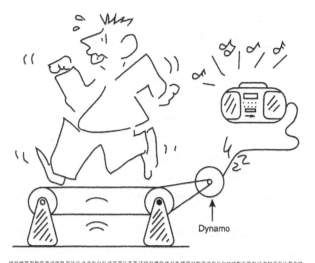

Figure 3.8.5 *Converting muscular energy into electrical energy.*

How to Build the Eolic Generator

Figure 3.8.6 shows the complete circuit for the eolic generator.

The capacitor is used as an energy reservoir, keeping the output voltage almost constant, even when the speed or power of the wind changes. The diode does not allow the current to return to discharge the motor across the coils. The larger the capacitor is, the more energy it can store, and the lower the voltage changes in the output will be when the speed of the wind varies.

The fan can be coupled to the motor directly. If gears or other mechanical systems are used, be sure that they don't cause energy losses, thus reducing the output of the generator.

Figure 3.8.7 shows a system using a small plastic fan being used in eolic demonstrations as the low-power energy source (i.e., as a replacement for the wind).

When mounting, it is necessary to determine the direction of the rotation relative to the diode. Rotation in one direction will cause the pole in the diode to be positive. And if the rotation is inverted, the pole in the diode will be negative, and no current will flow to the circuit.

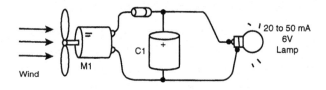

Figure 3.8.7 *A low-power eolic generator.*

Parts List—Eolic Generator

- M1 Small DC motor (3 to 12 volts)
- D1 1N4002 or equivalent silicon rectifier diode
- C1 1,000 μF to 4,700 μF × 6 volts or more electrolytic capacitor

Fan, wires, terminal strip, solder, etc.

Testing and Using

It is important to know the amount of energy produced by your eolic generator. If you have a multimeter, it is easy to test the generator. Adjust the multimeter to read DC volts on a low-voltage scale and connect it to the output of the generator, as shown in Figure 3.8.8.

The figure shows a low-cost analog multimeter, but you can make the same test using a digital multimeter. By turning the blades of the fan from any wind source (a small household electric fan, for instance), the generated power will move the multimeter's needle, indicating the voltage in the output. If no voltage is produced, invert the wires to the motor (the diode is reversed biased).

If you do not have a multimeter, you can use the circuit LED and lamp shown in the following paragraphs on powering circuits.

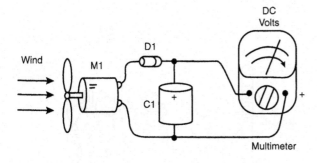

Figure 3.8.8 *Testing the generator with a multimeter.*

Powering Circuits

The amount of power generated by the eolic generator varies across a large range of values. Therefore, don't expect to power high-power devices using your prototype. With that in mind, some suggestions on low-power projects that can be powered by your project are also included.

LEDs and Lamps

Small LEDs and low-voltage lamps (such as the ones used in a flashlight) don't need large amounts of power to operate. Figure 3.8.9 shows how you can power those devices using an eolic generator.

The lamp is a low-power 6 V × 50 mA type. More than one LED can be powered from your generator. As you will observe, the amount of light will depend on the speed of the fan.

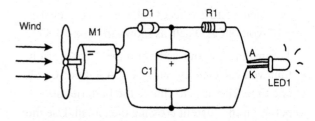

Figure 3.8.9 *Powering LEDs and lamps.*

Parts List—Powering Lamps and LEDs

L1   6 V × 50 mA incandescent small lamp

LED1 Any LED (red, green, yellow)

R1   1 kΩ × 1/8-watt resistor (brown, black, red)

Wires, solder, terminal strip, etc.

Battery Charger

The eolic generator can charge nickel cadmium (NiCAD) cells using the circuit shown in Figure 3.8.10.

This is the simplest version of a battery charger. The maximum number of batteries under charge depends on the voltage produced by your generator. To determine how many batteries you can charge at one time, it is necessary to measure the output voltage of your generator. A multimeter shown in the testing and using section can be used for this task.

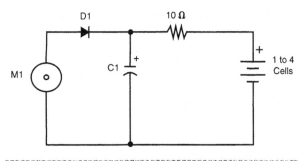

Figure 3.8.10 *NiCAD battery charger using the eolic generator.*

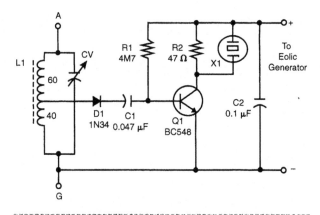

Figure 3.8.12 *Experimental AM radio power from the wind.*

For each 1.5 to 2 volts of voltage, you can charge one 1.2-volt NiCAD cell, as shown in Figure 3.8.11.

For instance, if your generator produces a 6-volt output, you can charge four AA, AAA, D, or C cells at one time. The time to charge the batteries also depends on the current crossing the circuit. Use the multimeter to measure the current.

If your generator is powerful enough to produce more than 9 volts, and the current rises up to 500 mA, you can add the constant current source as was introduced in Project 7. A constant current source will help you create a better battery charger.

by the circuit from a 2-volt source is less than 100 μA.

Figure 3.8.13 shows how the experimenter can build this radio using a terminal strip as a chassis. Of course, the radio can be mounted using other techniques, such as using a *printed circuit board* (PCB) or solderless board.

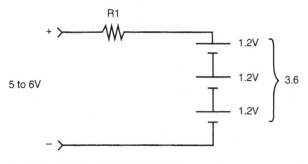

Figure 3.8.11 *You need 1.5 volts at least for each cell.*

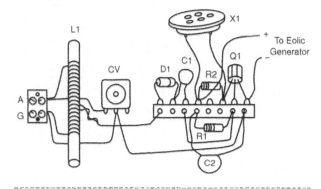

Figure 3.8.13 *The experimental radio mounted on a terminal strip.*

Experimental Radio

Another simple circuit that can be powered from your eolic generator or even from less powerful alternative energy sources is the experimental radio shown in Figure 3.8.12.

Even chemical and solar cells with voltages in the range of 1.0 to 6.0 volts can be used to power this very low-current circuit. In fact, the current drained

Only one transistor is used as a signal amplifier in this radio. Therefore, the radio is not overly sensitive and needs a long antenna to pick up local stations. The antenna can be made with a piece of long wire, 5 to 20 meters long. The ground connection is important and can be made through any metallic body in contact with the earth or even with your body. Even holding the ground alligator clip between your fingers will give good results. Remember to never use the radio during a thunderstorm!

L1 is formed by using 40 to 60 turns of 28 AWG enameled wire in a ferrite core with a length of 15 to 20 centimeters and a diameter of 1 to 1.5 centimeters. The variable capacitor can be taken from any old AM radio.

The transducer is made from a high-impedance piezoelectric phone such as the ones found in telephones, buzzers, and other appliances. Take care to not use a low-impedance type as it won't function in this project.

Parts List—Experimental Radio

Q1	BC548 or any *negative-positive-negative* (NPN) general-purpose transistor
D1	1N34 or 1N60 or any germanium diode
R1	4.7 MΩ × 1/8-watt resistor (yellow, violet, green)
R2	47 kΩ × 1/8-watt resistor (yellow, violet, orange)
C1	0.047 μF (47 nF) ceramic or polyester capacitor
C2	0.1 μF (100 nF) ceramic or polyester capacitor
X1	Piezoelectric transducer
L1	Antenna coil (see text)
CV	Variable capacitor (see text)
A, G	Antenna and ground connections

Ferrite core, terminal strip, antenna, alligator clip, wires, solder, knob for the variable capacitor, etc.

Automatic Light

The circuit shown in Figure 3.8.14 activates a small lamp at dusk and turns it off when the sun comes up.

The wind in the eolic generator charges the batteries. When the light on the sensor is cut, the batteries power the lamp. The lamp will remain on until the batteries discharge or the sun rises again.

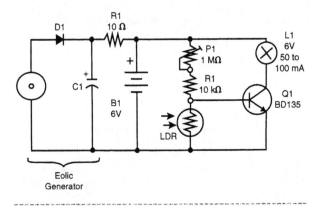

Figure 3.8.14 *Automatic light using the eolic generator.*

If the wind does not stop, the battery will continue to be in the charging process. The sensor must be installed inside a tube so that it receives only the ambient light. P1 allows you to adjust for the ideal level of light that will turn the light on and off.

Parts List—Automatic Light

Q1	BD135 or equivalent medium-power NPN transistor
D1	1N4002 silicon rectifier diode
P1	1 MΩ trimmer potentiometer
R1	10 Ω × 1-watt resistor (brown, black, black)
R2	10 kΩ × 1/8-watt resistor (brown, black, orange)
L1	6-volt × 50 or 100 mA lamp
LDR	Light-dependent resistor (any type)
B1	4 AA or D rechargeable nicad cells
X1	Eolic generator

Cell holder, plastic box, terminal strip, wires, solder, etc.

Anemometer

Because the voltage produced by the eolic generator is proportional to the speed of the wind, you can use this project as a simple anemometer, a device used to measure only the speed of the wind. Figure 3.8.15 shows how you can use your eolic generator as an anemometer.

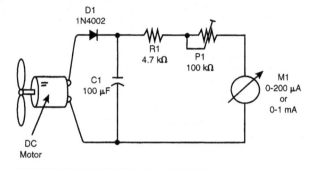

Figure 3.8.15 *The eolic generator as an anemometer.*

You can modify the system to measure the wind that is powering the motor by using small paper cups (see Figure 3.8.16).

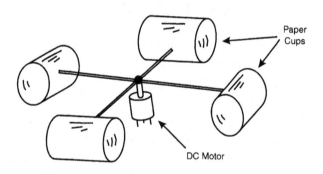

Figure 3.8.16 *Transforming the eolic generator to an anemometer by adding paper cups.*

The meter is an analog galvanometer or, if you prefer, one of the galvanometers suggested in Project 5 on experimenting with galvanometers.

P1 is adjusted according to a reference for the speed of the wind. The Beaufort scale, given in the following table, allows you to determine the approximate speed of the wind by its effects.

Force	Name	Description	Approximate Speed (mph)
Force 0	Complete calm	No motion. Smoke rises straight up.	Less than 1
Force 1	Light air	Smoke drifts.	1 to 3
Force 2	Light breeze	Wind felt on face. Leaves rustle. Weather is usually clear.	4 to 7
Force 3	Gentle breeze	Leaves and twigs move. Light flags flap.	8 to 12
Force 4	Moderate breeze	Small branches move.	13 to 18
Force 5	Fresh breeze	Bushes and small trees sway. Crests are common on the sea and known as "white horses."	19 to 24
Force 6	Strong breeze	Wind whistles in electricity and telephone wires. Hard to use umbrellas.	25 to 31
Force 7	Near gale	Whole trees sway and it becomes hard to walk in the wind. Sky may be dark and stormy.	32 to 38
Force 8	Gale	Now very difficult to walk and tree twigs start to break.	39 to 46
Force 9	Strong gale	Tiles and chimneys blown from roofs and branches may snap. Sky may be covered by clouds.	47 to 54
Force 10	Storm	Trees are uprooted and severe damage in buildings.	55 to 63
Force 11	Violent storm	Widespread damage is caused to buildings.	64 to 72
Force 12	Hurricane	Severe devastation.	Over 73

Parts List—Anemometer

D1 1N4002 silicon rectifier diode

C1 100 µF × 16-volt electrolytic capacitor

M1 0 to 200 µA or 0 to 1 mA analog meter

R1 4.7 kΩ × 1/8-watt resistor (yellow, violet, red)

P1 100 kΩ trimmer potentiometer

Plastic box, wires, solder, etc.

Cross Themes

Alternative energy sources are today being studied in many schools, beginning in elementary school science courses, all the way to physics and geography at the university level.

Both the most basic and the most complex versions (such as the one with a voltage regulator) of the eolic generator can be used when experimenting with a variety of alternative energy sources. It can be used to illustrate the following crossover themes:

- To show how the power of the wind can be converted into electrical energy
- To study the amount of energy that can be generated by a particular source
- To show how dynamos work
- To compare one energy source with other sources in terms of pollution, efficiency, and so on

Additional Circuits and Ideas

As explained previously, when building eolic generators you should not be limited to small DC motors only. Other generators with higher performance in energy conversion can be powered by wind energy.

In particular, we suggest a bicycle dynamo that comes in a variety of styles, starting with the unregulated units that are intended to power only lamps to the sophisticated units that have electronic circuits to charge the batteries of cellular phones. These bicycle dynamos need more speed and power to be used but can generate much more energy, 6 to 15 volts with a current of up to 2 amps in some cases. They are a good solution for the evil genius who wants a greater challenge beyond those given here.

Voltage Regulator

Sensitive electronic applications can be damaged if powered by a source presenting voltage changes, and this is exactly what our eolic generator can cause. Radios, calculators, and clocks are examples of voltage-sensitive devices. To power those devices, it is important to add a voltage regulator. Figure 3.8.17 shows two voltage regulators.

The first circuit is a low-power version recommended for use with devices pulling less than 5 mA. The second circuit can be used with devices pulling up to 1 amp.

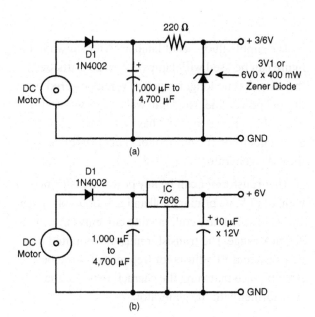

Figure 3.8.17 *Voltage regulators for the eolic generator.*

High-Voltage Inverter

If your eolic generator is powerful enough to source voltages between 5 and 12 volts with currents higher than 100 mA, you can power a fluorescent lamp using an inverter.

An inverter, such as the one shown in Figure 3.8.18, can convert low continuous voltages into high alternating voltages. The high voltage, when applied to a fluorescent lamp, can ionize the gas inside to make it glow.

Section Three   The Projects

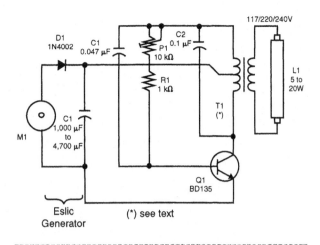

Figure 3.8.18 *A fluorescent lamp powered by an inverter.*

The simple fluorescent lamp inverter shown in the figure can be used with lamps of 5 to 20 watts, even the ones that no longer function when powered from the AC power line. The lamp will not glow with the original brightness but will have a reduced glow that is dependent on the amount of energy supplied by the eolic generator.

The transformer can be any power transformer with a 117 VAC primary coil and a 5- or 6-volt *center tap* (CT) secondary coil, rated to currents in the 50 to 300 mA range. The transistor must be mounted on a heatsink, and P1 adjusts for frequency for the best performance matching the characteristics of the transformer with the generator.

> **Caution!** Do not attempt to power any electronic device with this circuit. The output voltage is not a sine wave and the frequency is not 60 Hz.

Parts List—Inverter

Q1 BD135 or equivalent NPN medium-power transistor

T1 Transformer (see text)

R1 1 kΩ × 1/8-watt resistor (brown, black, red)

P1 10 kΩ trimmer potentiometer

C1 0.047 μF ceramic or polyester capacitor

C2 0.1 μF ceramic or polyester capacitor

L1 Fluorescent lamp, 5 to 20 watts

PCB or terminal strip, heatsink for the transistor, plastic box, wires, solder, etc.

The Technology Today

In many places around the world where wind is constant and intense, large eolic generators are used to power many kinds of electrical devices. Uses vary widely, from simple systems for pumping water from deep wells to major systems that light entire cities in the northwest of Brazil.

The basic eolic generator used today employs a large fan with three or more blades directly powering an alternator or dynamo (see Figure 3.8.19).

A "battery" of those generators can generate tens, hundreds, or thousands of kilowatts of electricity.

Figure 3.8.19 *Energy from the wind (Photo courtesy NASA).*

Project 9—Electronic Cannon

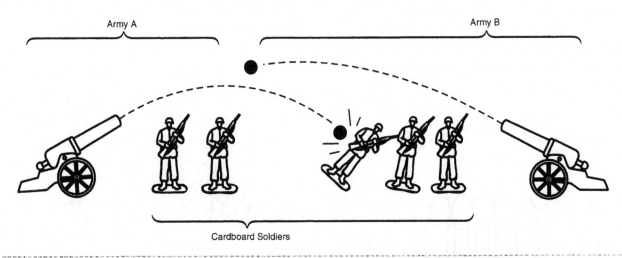

Figure 3.9.1 *Two electronic cannons. The armies are formed by small cardboard soldiers.*

Launching a cannon shell to distances up to many meters using the energy stored in a capacitor and converted in a magnetic field is the purpose of this project.

The electronic cannon can be added to your combat robot as an efficient weapon to destroy enemies in a robot war or it can be part of your army of stationary cardboard or plastic soldiers. Participate in a war involving two armies, one commanded by you and the other commanded by a friend, as Figure 3.9.1 suggests.

Of course, the shell is not heavy enough to injure anyone. It is a small piece of wood, plastic, or even a pea. The efficiency of the cannon and the distance it will fire depends only on the skill of the evil genius to mount the cannon.

In the basic version, shown in Figure 3.9.2, the cannon has power enough to launch a small shell to distances of some meters and can be used as a toy, in demonstrations, and even in classes to teach ballistics, physics, or mechanics.

The circuit can be built in several ways, according to the degree of difficulty the reader wants for the project.

Figure 3.9.2 *Basic cannon built using cardboard and wheels taken from a toy.*

Objectives

- Show how a solenoid works.
- Study energy conversion and how energy can be stored in a capacitor.
- Perform experiments with ballistics.
- Perform combat or war between armies using cannons as weapons (the winner will be the one who shod more enemy soldiers).
- Design a competition where the challenge is to build the cannon that can launch the shell the farthest distance.

Section Three   The Projects

How It Works

When an electric current flows across a wire, a magnetic field is created. We explored this idea in our electromagnet in Project 6. The magnetic field can be reinforced if the wire forms a coil, as shown by Figure 3.9.3.

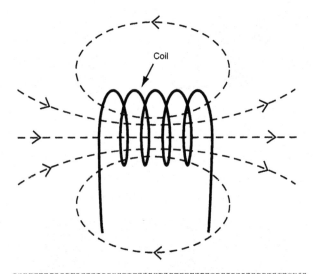

Figure 3.9.3 *In a solenoid, the magnetic field is strongest inside it.*

If a piece of any magnetic metal is placed near the coil, a magnetic force will attract the piece of metal when the current flows through the coil. This is the operating principle of a solenoid, as shown in Figure 3.9.4.

Solenoids are used to open doors, to move mechanical parts of many appliances such as CD players and DVDs, and to open and close water valves in clothes and dishwashers.

If the solenoid is strong enough to pull a small piece of metal through the core, it can launch an object a good distance, as shown by Figure 3.9.5.

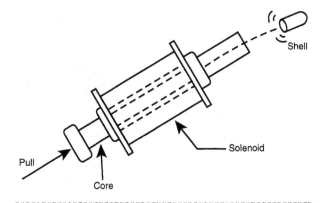

Figure 3.9.5 *Launching a cannon shell using a solenoid.*

This is the principle behind our magnetic cannon. A small solenoid receives current enough to pull the metal core very fast and with enough force to launch the cannon shell. Because the power of the cannon depends on the amount of current applied to the coil, a special circuit is used. The aim of this circuit is to generate a powerful current pulse using the energy stored in a capacitor.

This circuit is formed by a small transformer that generates a low AC voltage. This AC voltage is rectified by diodes and used to charge a large capacitor.

The larger the capacitor is, the more energy can be stored. By using capacitors in the range of 10,000 and 30,000 µF, much more energy can be stored than can be generated by small cells or other electric sources. This is true especially considering that the energy must be delivered very quickly. The main limitation to using capacitors as an energy source is that they can generate large amounts of energy but only at very slow speeds.

When the capacitor is connected to the coil, the discharge lasts only a few milliseconds, but an intense current will flow, producing a strong magnetic field. This field is enough to pull the metal core with such force to launch the shell, as shown by Figure 3.9.6.

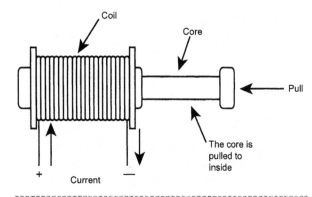

Figure 3.9.4 *Operating principle of a solenoid.*

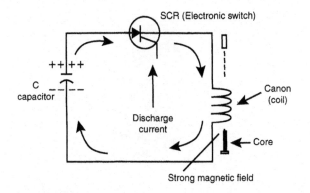

Figure 3.9.6 *The discharge of a capacitor across the cannon.*

Calculating the Power

It is easy to calculate how much power the capacitor can source when connected to the cannon. The amount of power stored in a capacitor is given by the following formula:

$$E = C \times V^2$$

where:

E = energy stored in joules (J)

C = the capacitance in farads (F)

V = the voltage across the capacitor (V)

For a 10,000 µF (10,000 × 10⁻⁶ F = 0.01 F) charged with 30 volts, the energy is

$$E = \tfrac{1}{2} \times 0.01 \times 30 \times 30$$

$$E = 0.01 \times 225 = 22.5 \text{ joules}$$

If the energy is delivered by the cannon in 0.1 seconds, the instant power of the gun is

$$P = 22.5/0.1 = 225 \text{ watts (or about 1/3 horsepower = 1 Horse Power)}!$$

Remember that power is energy released per time unit, so power in watts can also be expressed in joules per second (J/s).

Of course, the discharge time is very short but along a finite time because the wires used in the coil have some resistance. Imagine how powerful this small cannon can be!

Holding Large Currents

One problem to consider is how to handle the very high currents during the capacitor discharge using common components. First, to have such a fast discharge from the capacitor, the coil must have a very low resistance.

In our circuit, the capacitor charges with a voltage of about 36 volts (the peak of 24 VAC), and the coil of the cannon has only 0.8 ohms of resistance, as shown by the measures taken with a digital multimeter shown in Figure 3.9.7.

Figure 3.9.7 *The coil resistance must be as low as shown in the figure.*

If we consider only the resistance in ohms and not the inductance, when the capacitor is connected to the coil, the current can reach a peak of 40 amps or more. Common switches can't handle this level of current without overheating (and burning the contacts) in only a few uses.

So the solution adopted to control the discharge of the capacitor is the use of a *silicon controlled rectifier* (SCR). Common SCRs, such as the TIC106, operate as electronic switches and can control continuous currents up to 4 amps and can support very short peaks, like the ones produced by our circuit, of 100 amps.

Therefore, the idea in this project, as shown in Figure 3.9.8, is to use an SCR as the switch to trigger the cannon. This makes the use of low-power switches and even sensors (such as *light-dependent resistors* [LDRs]) for triggering the cannon suitable. You can

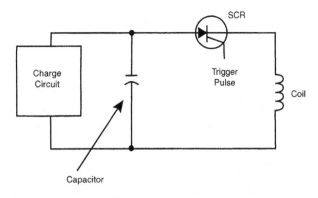

Figure 3.9.8 *An SCR is placed in series with the discharge circuit to act as a switch.*

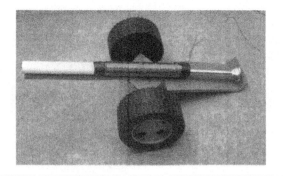

Figure 3.9.9 *The basic version (mechanical part).*

use a flashlight as a remote control to trigger your cannon.

Another problem to consider in the project is that, once triggered, the SCR remains in the on state, even after the trigger pulse disappears. To turn the SCR off, it is necessary to make the voltage across the circuit fall to zero for an instant. To do this, you need to disconnect the power. Even after the capacitor discharges, the voltage doesn't fall to zero on its own, due to the presence of the charging circuit. So it is necessary to disconnect this circuit before each time the cannon is used. A switch is added to keep the circuit connected during the time the capacitor is charging.

How to Build

The basic version of the electronic cannon makes a powerful weapon that can launch shells to distances of up to several meters, depending on the skills of the evil genius. A low-power version powered by cells will be explored too. As no electronic parts are needed for the low-power version, it is recommended for readers who are not as familiar with the mechatronic technologies used in the basic version.

The cannon can be mounted in several ways. In our basic version, we used a PVC or aluminum tube (1 cm in diameter $\times$ 12 cm long) mounted on a cardboard base with the dimensions shown in Figure 3.9.9.

The wheels were taken from a toy. However, depending on the application, the cannon can be fixed (avoiding the need for wheels). Remember that an inclination of about 15 degrees is necessary to move the core of the solenoid back to the shot position.

The shell for this version can be small plastic or wood balls or grains of dried beans, such as lentils or peas. The electronic circuit to drive the cannon is shown in Figure 3.9.10.

Any small transformer with a secondary rating for voltages of 9 and 12 volts, with a *center tap* (CT), and for currents in the range of 300 and 800 mA is suitable for this application. The primary coil must be rated according to your local power supply line, normally 117 VAC.

When mounting, observe the polarity of the diode and the capacitor. The value of the capacitor will determine the power of the cannon. We recommend large capacitors, such as the one found in the power supplies of computers, which range from 10,000 to 22,000 μF. The capacitor should be rated to voltages between 35 and 50 volts.

The small components of the circuit, such as the resistors, diodes, and SCR, can be soldered to a terminal strip. The final design of the circuit is shown in Figure 3.9.11.

The wires to the capacitor must be as short as possible, because the resistance caused by longer lines can result in energy losses. The placement of the components in a terminal strip and connections to the transformer and cannon are shown in Figure 3.9.12.

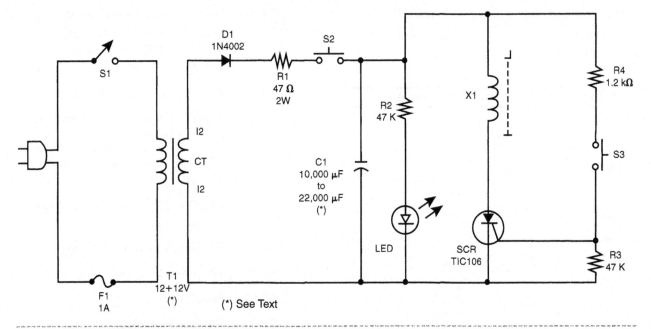

Figure 3.9.10 *Electronic circuit for the cannon.*

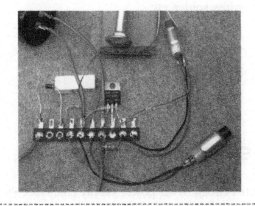

Figure 3.9.11 *Small components are soldered to a terminal strip.*

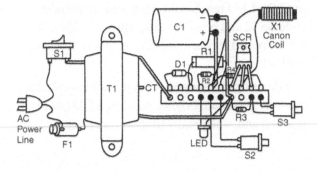

Figure 3.9.12 *The electronic circuit is housed inside a plastic or wooden box.*

The reader must take special care with the power cord and primary circuit of the transformer, as it is connected directly to the AC power line.

The Mechanical Part

The coil is calculated to present resistances of 0.7 to 2 ohms. This will keep the peak of the discharge current within the limits acceptable by the SCR (TIC106 or TIC116). In our prototype, we used 200 turns of 28 AWG enameled wire in a PVC tube, as shown in Figure 3.9.13. This results in a resistance of about 0.8 ohms.

It is easy to calculate how many turns (or the length) of a wire around a tube are required to maintain a specific resistance. Because the builder is free to change the characteristics of the project according to the components he or she has at hand, we recommend against following our instructions exactly. The evil genius's imagination should go to work here to

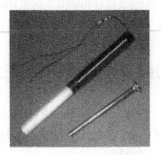

Figure 3.9.13 *The coil is formed by 200 turns of 28 AWG wire on a cardboard or aluminum tube.*

create a project with the best performance. When designing your coil, consider the following table of AWG resistances.

AWG Wire	Ohms per Kilometer
14	8.17
15	10.3
16	12.9
17	16.34
18	20.73
19	26.15
20	32.69
21	41.46
22	51.5
23	56.4
24	85.0
25	106.2
26	130.7
27	170.0
28	212.5

The size of the cannon, the length and diameter of the core, and the number of turns and wire used for the coil depend on many factors. The dimensions given in the text are for reference only. The reader is free and encouraged to make changes.

The cannon should be mounted on a cardboard or plastic base inclined 30 degrees to allow the core to slide back to the original position after each shot. Wheels were added to give the appearance of a real cannon. It is up to the reader to decide on the final appearance of the cannon, according the use. The reader can also use his or her evil genius imagination to add a mechanism that inserts a new shell after each shot.

Parts List—Electronic Cannon

SCR	TIC106 or TIC116 (B or D) SCR
D1, D2	1N4002 or equivalent silicon rectifier diode
LED	Common *light-emitting diode* (LED) (any color)
T1	Transformer: primary coil according to the AC power line, secondary coil to 12 volts CT, and 300 mA to 500 mA
R1	47 Ω × 2-watt wire-wound resistor
R2, R3	47 kΩ × 1/8-watt resistor (yellow, violet, orange)
R4	1.2 kΩ × 1/8-watt resistor (brown, red, red)
C1	10,000 to 22,000 μF × 35-volt electrolytic capacitor
X1	Cannon (solenoid) (see text)
S1	SPST switch (on/off)
S2, S3	Pushbutton
F1	1 A fuse and holder

Terminal strip, power cord, enameled wire, wires, solder, etc.

Material for the cannon: cardboard, PVC tube, plastic wheels, metal core (screw), enameled wire, etc.

Testing and Using

Plug the circuit to the power supply line. Place the cannon in a shooting position and insert the shell, keeping the core in the correct position. Switch S1 on to power the circuit. To charge the capacitor, press S2 until the LED reaches the maximum brightness. Release S2 and press the pushbutton S3. The core will be pulled forward inside the cannon, launching the shell.

Preparing for a new shot is simple: Press S2 again until the maximum brightness of the LED has been reached and, when ready to shoot, press S3.

Exploring the Project

The cannon is more than simply a fun, curious project. It can be used to teach theories behind ballistics or as an experimental tool in physics classes. It can even be used as a demonstration in a history class.

The following experiments will help the reader explore the project further.

Studying Ballistics

This experiment can be performed as a crossover theme in courses on physics (high-school-level curriculum). Place the cannon at different angles, shooting and measuring the distance reached by the shell (see Figure 3.9.14).

If the mass of the shell is known, the place where the shell falls and the angle of the shot can be used to calculate the initial speed of the bullet (Vo) and the kinetic energy. The experiment can performed after theoretical classes about ballistics.

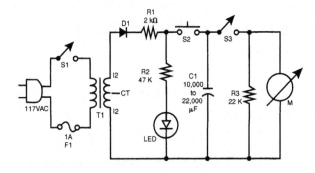

Figure 3.9.15 *Charging and discharging a capacitor to determine the RC constant of a circuit.*

Figure 3.9.15. The idea is to charge the capacitor by a predetermined voltage (measured by the multimeter) and then trace the discharge curve when the capacitor is connected to a known resistor.

The graphs obtained during this experiment allow the reader to show the calculation of the energy stored in the capacitor and the time constant of an RC circuit.

To obtain the points on the graph, charge the capacitor by pressing S2. Then close S3 and measure the voltage in constant time intervals (every 20 seconds, for instance), placing the voltage measurements in a table. When you plot the results, you will have the discharge curve of the capacitor.

The resistor can be replaced by a 24 V × 50 mA lamp. If you intend to use a 12-volt lamp, you must add a resistor in series to reduce the applied voltage. Remember that the capacitor charges with about 35 volts in this circuit. Ohm's law can be used to calculate the series resistor.

Another possibility is to use an LED in series with a 22 kΩ × 1/4 W resistor as a discharge circuit. Remember that the LED is not a linear load when the voltage falls above 2 volts.

Observe that the LED that indicates a full charge is placed before S2 in series. This is necessary because, after S2 is released, if the LED is placed after the switch, as in the cannon, the capacitor will discharge slowly across the LED.

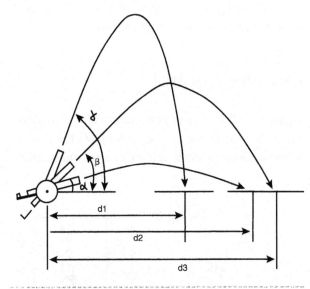

Figure 3.9.14 *Show how the angle determines the range of the cannon.*

Measuring the Charge Stored in a Capacitor and the Time Constant of a Remote Control Circuit

Another experiment that can be performed using the charging circuit of the cannon is the one shown in

Parts List—Measuring the Charge

- D1 — 1N4004 or equivalent silicon rectifier diode
- LED — Common LED (any color)
- R1 — 2.2 kΩ × 1-watt resistor (red, red, red)
- R2 — 47 kΩ × 1/8-watt resistor (yellow, violet, orange)
- R3 — 22 kΩ × 1/2-watt resistor (red, red, orange)
- S1, S3 — SPST on/off switches
- S2 — SPST pushbutton
- F1 — 1-amp fuse and holder
- T1 — Transformer: primary coil to 117 VAC or according to the local power line, secondary coil to 12 V CT, and 300 mA or higher
- C1 — 10,000 μF to 22,000 μF × 35-volt electrolytic capacitor
- M — Common analog or digital multimeter (5,000 Ω/V or more sensitive)

Power cord, terminal strip, wires, solder, etc.

A Battle Using Electronic Cannons

A battle between two armies will test your skills with the use of your electronic cannons. The armies are made up of plastic or cardboard soldiers as shown in Figure 3.9.16.

You can add fortifications, military vehicles, and so on to create the look of a real battlefield. The battle can be fought in two ways: The winner can be the first to knock out all the opposing soldiers or the one who shoots the most soldiers in a predetermined time interval (30 minutes, for instance).

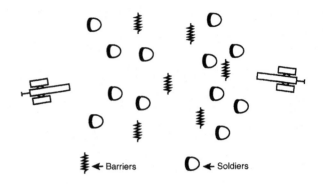

Figure 3.9.16 *Two armies in battle using electronic cannons.*

Additional Circuits and Ideas

The evil genius can create many other circuits starting from the basic project described here. In the next section we will give some suggestions.

Remote Control

Figure 3.9.17 shows a remote control circuit used to trigger the electronic cannon. This allows the reader to use a flashlight as the remote control transmitter.

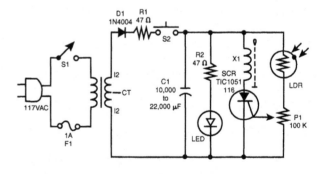

Figure 3.9.17 *The cannon is controlled by a flashlight with this remote control.*

The sensor is an LDR placed inside a cardboard tube to avoid any interference from ambient light in its operation.

P1 is a trimmer potentiometer used to make adjustments in sensitivity. The tube with the sensor must be positioned in the direction of the light used for the remote control.

The remote control is used only to trigger the circuit. As we learned earlier, S2 must be pressed again after each shot in order to recharge the capacitor. You can create a circuit to trigger a relay using the light beam, replacing S2 with the contacts of the relays. Therefore, by using two sensors, you can point the flashlight at one of them to charge the capacitor and at the other to trigger the cannon. The next step, of course, is the creation of a mechanism to insert a new bullet.

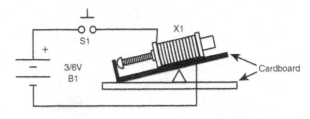

Figure 3.9.18 *Low-power cannon using D cells as the power supply.*

Parts List—Remote Control

SCR	TIC106 or TIC116 (B or D) SCR
D1, D2	1N4004 silicon rectifier diode
LED	Common LED (any color)
T1	Transformer: primary coil to 117 VAC, secondary coil 12 volts CT, and 300 mA or more
R1	47 Ω × 2-watt wire-wound resistor
R2	47 kΩ × 1/8-watt resistor (yellow, violet, orange)
P1	100 kΩ trimmer potentiometer
LDR	LDR of any type, such as a *cadmium sulfide* (CdS) cell
C1	10,000 to 22,000 μF × 35-volt electrolytic capacitor
S1	SPST on/off switch
S2	SPST pushbutton
F1	1-amp fuse and holder
X1	Electronic cannon (see text)

Terminal strip, wires, power cord, solder, etc.

Low-Power Version

A very low-power version of the electronic cannon without the need for a special drive circuit is shown in Figure 3.9.18.

This circuit can be powered directly from two or four D cells. The shell can be launched to distances up to some meters according to the skills of the builder. Reducing friction and using a very light shell will ensure better performance.

The coil is formed by 200 to 500 turns of 28 to 32 AWG wire around a cardboard or plastic tube with a diameter of approximately 0.5 centimeters and a length of 5 centimeters. The core uses a 2- × 1/8-inch screw.

The format is the same as in the basic version. The cannon must be mounted on an incline to allow the core to slide back to the firing position after each shot.

Do not keep S1 pressed after the cannon is triggered. This will help maintain the charge in your batteries, as the current drains quickly when S1 is pressed.

Parts List—Low-Power Version

X1	Cannon (see text)
B1	2 to 4 D cells with holder
S1	SPST

Wires, enameled wire, solder, etc.

Building a Catapult

Another antique weapon can be built based on the same principle of the electronic cannon: a catapult. As Figure 3.9.19 suggests, the core of a solenoid, when pulled back quickly, can launch a stone. Of course, the size of the stone (you can use a dried bean as the stone in a miniature project) and the distance fired will depend on the power of the solenoid.

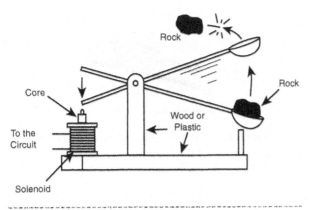

Figure 3.9.19 *A catapult using a solenoid.*

You can build two versions of this weapon: a large size or high-power version (based on the plans for the basic version of the cannon) or a low-power version with circuits powered from D cells. The critical points to be considered when developing the catapult version of the cannon is to ensure that the catapult returns to the firing position after each shot and that the launch of the stone is in the correct direction.

Developing a Supercannon

The idea of a powerful cannon has been explored by some evil geniuses, using a sequential trigger and very large capacitors. This is not something we recommend building. We do, however, recommend using this idea as an exercise simply to analyze the possibilities.

A common design of a supercannon places several magnetic rings (electromagnets) across a PVC transparent tube, as shown in Figure 3.9.20.

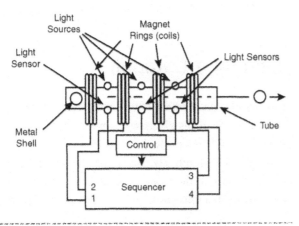

Figure 3.9.20 *Basic idea of a super electronic cannon.*

The shell is made of a piece of magnetic metal that can be pulled quickly through the cylinder by the rings when they are energized by a current. When triggered, the first bank of capacitors discharges across the first ring, pulling the shell and giving it its first impulse. As long the shell passes the first ring, it crosses a light beam triggering the second impulse circuit.

This circuit is another bank of capacitors that discharge across the second magnetic ring. A new pull is produced, adding impulse to the shell and increasing its speed.

Several rings are used to give to the shell fantastic speed. The shell is finally launched with enough power to destroy what it hits. Experimental projects have resulted in shells that can put a hole in a cement wall.

> **Caution!** Do not try this version without an adult to give you the necessary support!

The Technology Today

Solenoids are an important part of many devices. Cars, VCRs, DVD players and recorders, CD players and recorders, dishwashers, washing machines, and many other appliances use solenoids.

The solenoids found in these appliances can be small enough to fit in the palm of your hand or large enough to weigh several pounds. Very thin wires and small parts are used to build solenoids that ultimately release great mechanical power when commanded.

In your car, solenoids have many applications. The simplest of which is to release the lock of the trunk when you press a button on the panel.

Ideas to Explore

Use the following ideas to increase the performance of your cannon or to simply learn more about the project:

- Add mechanical resources to slide back the solenoid core after each shot.

- Design a circuit to turn on an alarm (see the combat robot for circuits and oscillators that can be used as alarms) when the capacitor is fully charged.
- Create a method to launch an arrow instead of a shell, such as an old-fashioned crossbow.
- Explore the Internet to find projects related to Gauss cannons.
- Find applications in the home where solenoids are used. Try to design and build a solenoid project for the home. Many solenoids are powerful and can be powered from 6 or 12 volts. Dishwashers and washing machines also use powerful solenoids, but they must be powered from the AC current. You can perform experiments with them, but take care to not touch any part of their circuitry.

Project 10—Experiments with Lissajous Figures Generated by a Laser

The purpose of this project is to generate some Lissajous figures on a screen, wall, or other clear surface using a laser beam. The nice forms of the Lissajous figures and the change of their shapes, size, and other characteristics allow their use in many interesting applications such as in decoration, science fairs, demonstrations, and even to teach physics.

The project described in the basic version is very simple and uses a low-cost laser pointer or laser modules that can be powered from cells. Of course, the reader is free to follow the suggestions to upgrade the project.

The project uses an electronic circuit to control the movements of the light beam (generating the signals that result in Lissajous figures) and a mechanical part to move mirrors. Therefore, it can be considered a mechatronic project. Figure 3.10.1 shows some patterns that can be produced by this project using a simple laser pointer.

In the basic version of this project, we will use common parts from everyday items and some basic electronic components. This makes it ideal for the reader who is not experienced with advanced technologies or has only basic mechanical and electronic parts to work with.

Figure 3.10.1 *Lissajous figures produced by a laser pointer in this project.*

- Build a Lissajous figure generator using a laser pointer.
- Perform demonstrations in physics classes.
- Create a decorative apparatus to be used in fairs, events, stores, parties, and so on.
- Know how Lissajous figures are used to measure the frequency of a signal.

Objectives

- Learn what Lissajous figures are and how they can be produced.

What Lissajous Figures Are

Any periodic movement, such as an oscillating pendulum, can be described by a trigonometric equation. Projecting the movement we will find a trajectory as shown by Figure 3.10.2.

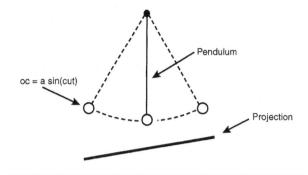

Figure 3.10.2 *The trajectory of an oscillating pendulum.*

As the reader can see, the movement occurs in a straight line, because it is unidimensional. This is the simplest oscillating movement, but not the only one we can find in nature.

What happens if two oscillating movements are combined? Imagine a small object attached to four springs, as Figure 3.10.3 shows.

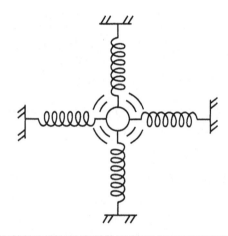

Figure 3.10.3 *A vibrating object attached to springs.*

In the arrangement shown by the figure, the object can oscillate in any direction. A pen can be used to draw the projection of the movement on a piece of paper.

If the object vibrates in alignment with the X axis, the figure will be a straight line in the X axis. The same occurs if the object vibrates in alignment with the Y axis. But what will happen if the oscillations are in the X and Y axes at the same time?

As Figure 3.10.4 shows, the trajectory of the object projection on paper depends on several factors:

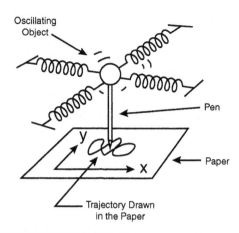

Figure 3.10.4 *Trajectory of the oscillating object drawn on paper.*

- The frequency of the oscillations in both axes
- The amplitude of the oscillations
- The function that produces the vibrations

The French mathematician Jules Antoine Lissajous worked in the nineteenth century studying vibrating objects and, in particular, sound waves. To perform his experiments, a mirror was attached to a tuning fork. A beam of light was focused and reflected by the mirror. The reflected beam was then directed to another mirror attached to another tuning fork in a different position. The two tuning forks were placed in a position so that they would vibrate perpendicular to one another, as shown in Figure 3.10.5.

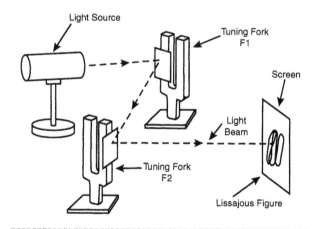

Figure 3.10.5 *The experiment of Lissajous.*

The beam resulting from the second reflection was then directed to a screen where interesting figures appeared. He found that when the frequencies of the tuning forks had a predetermined ratio of values, the figures had special shapes. He also noticed that the figures appeared particularly when the frequency ratio of the vibrations could be represented by integers.

Those figures are now called Lissajous figures and beyond their value as a curiosity they have some practical utilities. Figure 3.10.6 shows some Lissajous figures and the corresponding frequency ratio of the vibrations.

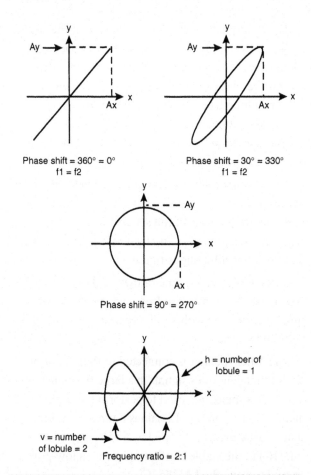

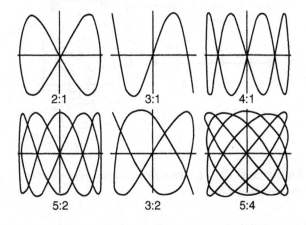

Figure 3.10.6 *Some Lissajous figures and the corresponding ratio.*

Figure 3.10.7 *Phase shift and amplitude ratio measured by a Lissajous figure.*

Besides the information about the frequency ratio of two vibrations, or oscillations, Lissajous figures can provide additional information regarding the phase shift and amplitude ratio of the vibrations. Figure 3.10.7 shows how a Lissajous figure can provide that information.

It is important to see that sound waves resulting in Lissajous figures must be natural vibrations. These vibrations can be represented by the following equations:

$x = A1 \sin(\Omega 1.t)$

$y = A2 \sin(\Omega 2.t)$

The Project

Our basic version of this project is a simple Lissajous figures projector using a laser pointer as a light source and two loudspeakers in a mechanical vibrating system, coupled to a small mirror, as shown by Figure 3.10.8.

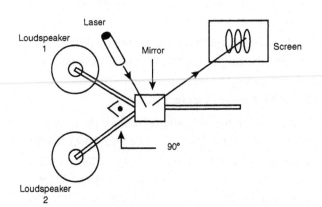

Figure 3.10.8 *How loudspeakers can be used to generate Lissajous figures.*

The loudspeakers are positioned so that the sound waves (movements) are perpendicular to each other when they reach the mirror. The result is that the mirror vibrates with a composite movement, a movement resulting from the oscillations of both speakers. Then the light beam of a laser pointer, which is focused on the same mirror, is reflected at angles that depend on the instantaneous position of the mirror.

As the mirror vibrates, the reflected beam draws a figure that is a direct result of the movement. If the vibrating frequencies of the speaker are in a ratio represented by integers, the figure projected on the screen will be a Lissajous figure.

In the basic project, one of the speakers is powered by an AC power line frequency of 60 Hz. A small transformer with a potentiometer to control the amplitude is used.

The other speaker is connected to the output of a circuit that generates signals in a large frequency range. This speaker is used to perform the experiments and to generate Lissajous figures of several shapes. This circuit is formed by a *555 integrated circuit* (IC) in the stable configuration and a low-power amplifier using the LM386 IC.

Although the 555 produces square waves, the resonance characteristics of the loudspeaker when vibrating produce quasi-sinusoidal vibrations. Therefore, the figures are not perfect. They have some distortion but with a pattern that resembles the original Lissajous figures.

Thus, it is enough to excite the audio amplifier and the loudspeaker in a way that results in acceptable Lissajous figures. An alternative to having perfect figures is to replace the square oscillator with an external signal generator or functional generator that can produce good sinusoidal waves.

By changing the frequency of the oscillator, you can find the frequencies that result in the best Lissajous figures. The block diagram for the complete circuit is shown in Figure 3.10.9.

As an optional laser source, replace the laser pointer with a small laser module and power it with a derived voltage source, such as the one from the main source used for the oscillators and amplifier. Figure 3.10.10 shows how a laser pointer can be adapted to be powered by an external power source.

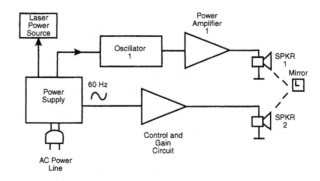

Figures 3.10.9 *Block diagram for the project.*

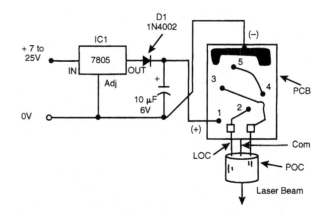

Figure 3.10.10 *Laser power source derived from the transformer.*

Laser pointers normally are powered from three cells (4.5 volts), but the nominal voltage for small modules can range from 4.5 to 6 volts. In our project, a 5-volt positive regulator was used to drive the laser module.

Building the Circuit

Figure 3.10.11 shows the electronic circuit for the Lissajous laser.

The electronic circuit can be built using a *printed circuit board* (PCB) or, for experimental purposes, a solderless board, such as the one shown in Figure 3.10.12. Of course, the evil genius is free to choose the mounting process that best suits the purpose of the project.

The transformer has a primary coil rated to the AC power supply line, usually 117 VAC. The secondary coil is rated to 6 volts with currents ranging from

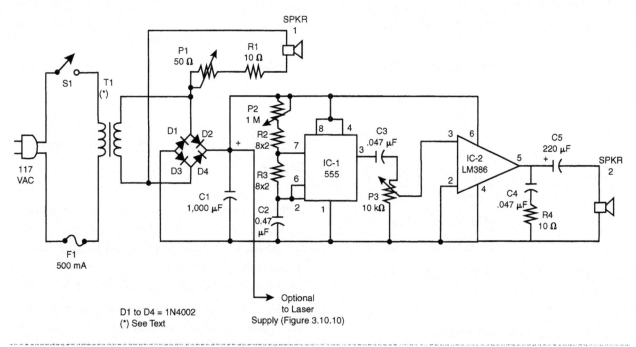

Figure 3.10.11 *Electronic circuit for the Lissajous laser.*

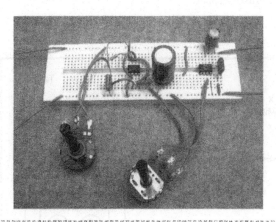

Figure 3.10.12 *Electronic circuit mounted on a solderless board.*

500 to 800 mA. The speakers are common types, with cone diameters ranging between 4 and 5 inches (10 and 12.5 centimeters) and impedances of 4 or 8 ohms. Larger speakers can also be tested, but require making adaptations to the mechanical system. Larger loudspeakers have the best performance because their cones transfer larger amplitude vibrations to the mirror.

The position of polarized components such as electrolytic capacitors, diodes, and ICs must be observed. P1 is a wire-wound potentiometer because it acts as a rheostat directly controlling the signal applied to the speaker. Don't use the electronic potentiometer described in this book because it is suitable only for motor and solenoid control.

Parts List—The Circuit

IC-1	555 IC timer
LM386	IC audio amplifier
D1, D2, D3, D4	1N4002 silicon rectifier diodes
SPKR1, SPKR2	Small loudspeakers (12.5 cm × 8 Ω)
T1	Transformer: primary coil to 117 VAC, secondary coil to 6 volts × 800 mA (see text)
C1	1,000 μF × 12-volt electrolytic capacitor
C2, C3, C4	0.047 μF ceramic or polyester capacitors
C5	220 μF × 12-volt electrolytic capacitor
P1	50 Ω wire-wound potentiometer
P2	1 MΩ linear or logarithmic potentiometer

P3	10 kΩ logarithmic potentiometer
S1	SPST switch (on/off)
R1	10 Ω × 2-watt resistor (brown, black, black)
R2, R3	8.2 kΩ × 1/8-watt resistor (gray, red, red)
R7	10 Ω × 1/8-watt resistor (brown, black, black)
S1	SPST on/off switch
F1	500 mA fuse and holder
C1	10 μF × 12-volt electrolytic capacitor
X1	Semiconductor laser module

PCB or solderless board, power cord, plastic box, wires, solder, knobs for the potentiometers, etc.

Optional for the laser supply:

IC1	7805 IC
D1	1N4002 silicon rectifier diode

Building the Mechanical Part

In the basic version, a piece of wood is used to keep the loudspeakers and the vibrating system in the correct position. As shown in Figure 3.10.13, the loudspeakers are glued to the base in the position determined by the arms where the mirror is attached.

The arm was made using bare wire 14 or 16 AWG. Three pieces of wire were soldered to form a Y. The reader must take care to form right angles between the two parts attached to the loudspeakers, as shown in Figure 3.10.14.

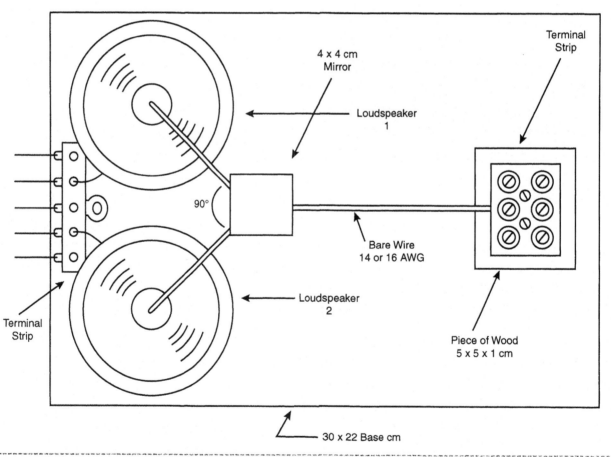

Figure 3.10.13 *The mechanical system.*

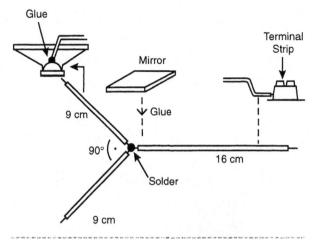

Figure 3.10.14 *The arms attached to the loudspeakers must be at a right angle and form a Y.*

The Y is kept in position by using a terminal strip and screws. The 4 × 4 centimeter mirror is glued to the center part of the Y.

To connect the system to the external circuit, a terminal strip can be used. Figure 3.10.15 shows the mechanical part of the project assembled and mounted on a wooden base.

For experimental purposes, the operator can hold a laser pointer and focus it on the mirror to create a reflected image on the wall or other surface. But to have a more stable image, the laser pointer can be fixed in a support.

Of course, in the final version, all the electronic parts can be housed inside a box, and the laser can even be fixed and powered by the same circuit.

Figure 3.10.15 *The loudspeakers and mirror are placed on a wooden base.*

Testing and Using

Set P3 to the minimum, cutting all the vibrations in the amplifier channel. Plug in the power cord to the AC power line. While focusing the laser pointer on the mirror, adjust P1 to produce an image on a screen like the one shown in Figure 3.10.16.

Now slowly adjust P2 and P3 so the vibrations are in the other channel. P3 adjusts the amplitude and P2 adjusts the frequency. Adjust P3 until you find a frequency that results in Lissajous figures.

If serious distortion occurs, change the loudspeakers so they are positioned with a slight inclination. Distortion is common because the vibrating system is not perfect and sound waves or secondary vibrations producing resonance can be added to the system. Extending the length of the mechanical arms where the mirror is placed may increase the level of performance as well.

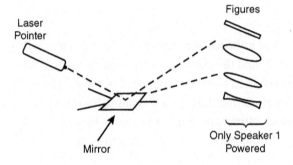

Figure 3.10.16 *Image produced when only SPKR1 is vibrating.*

Exploring the Project

Many methods exist for controlling a laser beam and producing good Lissajous figures. Our basic project used simple vibrating objects and a laser pointer. Of course, the evil genius may want to create other configurations.

Section Three   The Projects

Rotating Mirror

One of the simplest ways to create an oscillating reflective surface is by coupling mirrors to the shaft of small DC motors, as shown in Figure 3.10.17. Two small motors mounted at right angles to each other can be used to produce Lissajous figures.

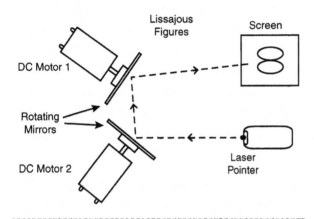

Figure 3.10.17 *Using two DC motors to rotate mirrors.*

The *pulse width modulation* (PWM) control (Project 3) and the electronic potentiometer (Project 7) can be used to change the speed of the motors and therefore find the mirrors' rotating frequencies, which ultimately result in Lissajous figures. Figure 3.10.18 shows how the controls can be wired to the motors.

The same power supply that sends energy to the motor can be used to power the laser by using a voltage step-down stage, as shown in the basic project.

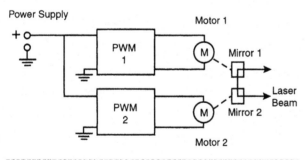

Figure 3.10.18 *Controlling the speed of the motors.*

Solenoids

Small solenoids are good sources of vibrations when they are coupled to mirrors, as shown in Figure 3.10.19.

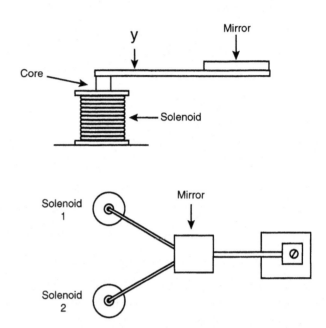

Figure 3.10.19 *A modulation system made with solenoids.*

This kind of circuit can be coupled directly to the output of the transformer (using a rheostat to limit the current) or to the output of oscillators and audio amplifiers.

The reader can perform experiments using solenoids built with 100 to 500 turns of 28 to 32 AWG wires around small cardboard or plastic forms.

Stepper Motors

Stepper motors are suitable for laser modulation and, in this case, can be controlled quite accurately. A 400-steps-per-rotation stepper motor offers an accuracy of better than one degree. See Project 14 for more information about stepper motors.

Another point to be considered when using stepper motors is that they can be controlled directly from a computer. In Project 14, the reader will find a circuit to control stepper motors that can be used to produce Lissajous figures.

Vibrating Blades

Figure 3.10.20 shows an interesting arrangement of vibrating blades driven by electromagnets.

The blades are made with ferrous materials that can be attracted by the magnet field produced by the electromagnets. The electromagnets in a simple version can be made with 100 to 200 turns of 28 to 32 AWG around a 1/2 × 1/8-inch screw.

The circuit can be driven by a small transformer (60 Hz) and any amplifier coupled to an oscillator or function generator. Notice that the vibrating blades must be mounted at right angles. Remember that the size of the figure depends on the power applied to the system.

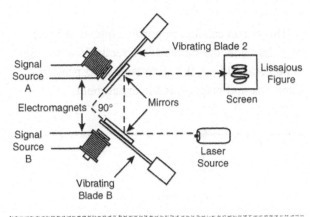

Figure 3.10.20 *Vibrating blades can also be used to control a laser beam.*

Computer Control

Multimedia audio amplifiers have two channels and use two small loudspeakers, as shown in Figure 3.10.21.

The reader can build an inexpensive system using loudspeakers and mount them in the same arrangement shown in our basic project. In this case, the frequencies to be produced by the speakers that will modulate the laser beam can be generated by soft-

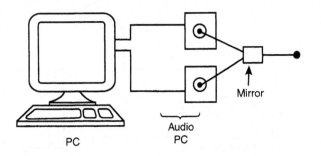

Figure 3.10.21 *Multimedia audio amplifiers.*

ware. The reader can add the program controls to adjust the exact points where the Lissajous figures are produced. Because the software makes it possible to generate exact frequencies, it is much easier to find the ratios that produce the figures.

Using Audio or Function Generators

Accurate sinusoidal waves can be generated using function generators or audio frequency generators found in electronics labs. Those instruments can be coupled to audio amplifiers to modulate the laser beam as shown in Figure 3.10.22. Any distortion in the figures will be the result of the mechanical modulation system.

Cross Themes

Lissajous figures are studied in the mechanics portion of a physics course at the high school level. The device can be used to make practical demonstrations regarding the generation of the curves.

Many activities related with the theme can be examined, such as the following:

- Determining the frequency ratio based on the figures
- Measuring the phase shift of two signals
- Analyzing the distortions and their causes
- Knowing how one frequency (60 Hz) determines the other (produced by the oscillator)

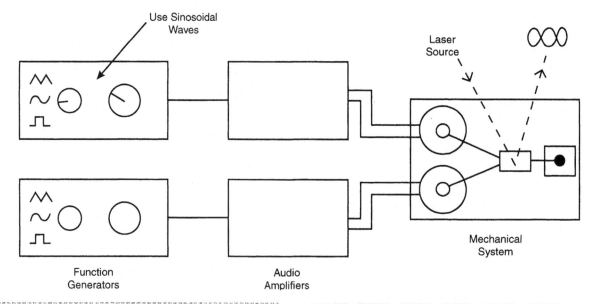

Figure 3.10.22 *Using audio or function generators.*

Additional Circuits

Oscillators suitable for this project and others are discussed in the following section.

Oscillator Using the 4093 IC

The circuit shown in Figure 3.10.23 produces a variable frequency signal. The Resistor Capacitor filter on the output smoothes the wave, changing its shape to a quasi-sinusoidal form.

This circuit can accept power supplies of 5 to 12 volts. Its signal can be used to directly drive the input of audio amplifiers, thus activating loudspeakers, solenoids, and magnets in the experiment.

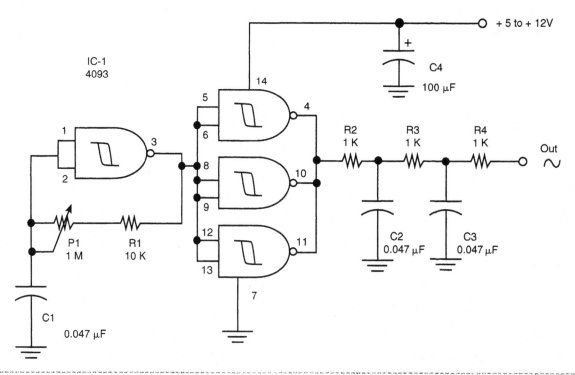

Figure 3.10.23 *Oscillator using the 4093.*

Parts List—Oscillator Using the 4093 IC

IC-1	4093 *complementary metal oxide semiconductor* (CMOS) IC
R1	10 kΩ × 1/8-watt resistor (brown, black, orange)
R2, R3, R4	1 kΩ × 1/8-watt resistor (brown, black, red)
P1	1 MΩ linear or logarithmic potentiometer
C1, C2, C3	0,047 µF ceramic or polyester capacitor
C4	100 µF × 12-volt electrolytic capacitor

PCB or solderless board, wires, etc.

Sine Wave Oscillator

Figure 3.10.24 shows an audio oscillator that produces a good low-frequency sine wave.

In a double-T oscillator, the frequency is determined by the capacitors and resistors in the T feedback network. The values of the components in the T network must have a ratio of values as indicated.

The potentiometer allows a change of frequency across a low range of values. When using this circuit to modulate the laser beam, you might have to add a switch to change the T networks according to the Lissajous figure to be produced.

The circuit can be powered by voltages as low as 6 volts, but the signal will be weak and need to be amplified by an external audio amplifier.

Parts List—Sine Wave Oscillator

Q1	BC548 or equivalent general-purpose *negative-positive-negative* (NPN) silicon transistor
R1, R2	100 kΩ × 1/8-watt resistor (brown, black, yellow)
R3	5.6 kΩ × 1/8-watt resistor (green, blue, red)
P1	47 kΩ linear potentiometer

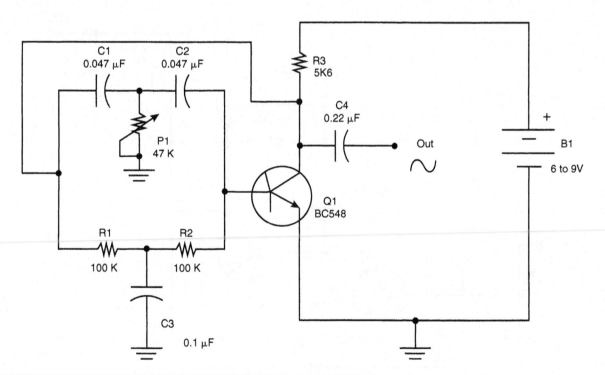

Figure 3.10.24 *Double-T sine wave oscillator.*

C1, C2   0.047 μF ceramic or polyester capacitors

C3   0.1 μF ceramic or polyester capacitor

C4   0.22 μF ceramic or polyester capacitor

B1   6- to 9-volt battery or cells

PCB or solderless board, cell holder or battery connector, wires, etc.

Transistor Audio Amplifier

Figure 3.10.25 shows an alternative audio amplifier used to drive the small loudspeaker from the signal coming from an audio oscillator.

The circuit can be powered by 3 to 6 volts. The circuit can be mounted on a small PCB or a solderless board for experimenting.

Alternative transistors can be used, and all other components are not critical. The amplifier will provide some tens of milliwatts to the loudspeakers, enough to make them vibrate, and therefore show oscillations in the mirror.

Parts List—Transistor Audio Amplifier

Q1, Q2   BC548 or equivalent NPN general-purpose silicon transistor

Q3   BC558 or equivalent *positive-negative-positive* (PNP) general-purpose silicon transistor

D1, D2   1N914 or 1N4148 general-purpose silicon diodes

R1   1 kΩ × 1/8-watt resistor (brown, black, red)

R2   47 kΩ × 1/8-watt resistor (yellow, violet, orange)

P1   10 kΩ logarithmic potentiometer

C1   10 μF × 12-volt electrolytic capacitor

C2   470 μF × 12-volt electrolytic capacitor

C3   1,000 μF × 12-volt electrolytic capacitor

6-volt power supply (4 AA cells), PCB or solderless board, wires, solder, etc.

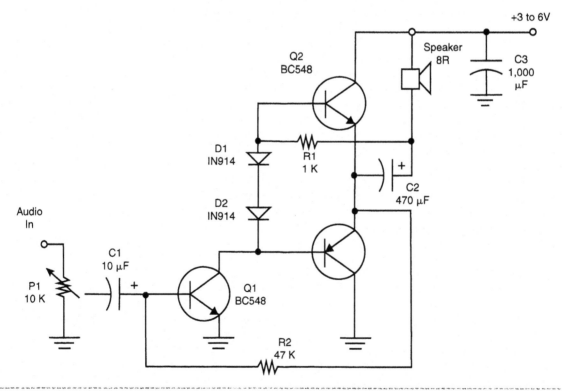

Figure 3.10.25   *Audio amplifier using a common bipolar transistor.*

High-Power Amplifier

For the reader who intends to operate a high-power laser such as a HeNe laser and needs a circuit that can drive large mirrors (10 × 10 centimeters, for instance), a powerful amplifier and large loudspeakers are needed. The circuit shown in Figure 3.10.26 can source about 8 watts when powered from a 13.2-volt power supply.

The ratio between R1 and R2 determines the voltage gain of the circuit. The circuit can be mounted on a PCB.

A heatsink must be attached to the IC. The power supply must be rated to 3 amps or more. Figure 3.10.27 shows a suitable power supply for this amplifier.

Any transformer with a primary coil according to the AC power line voltage and a secondary coil between 9 and 12 volts can be used.

Parts List—High-Power Amplifier

IC-1   TDA2002 IC audio power amplifier

R1   220 Ω × 1/8-watt resistor (red, red, brown)

R2   2.2 Ω × 1/8-watt resistor (red, red, gold)

R3   1 Ω × 1/8-watt resistor (brown, black, gold)

P1   10 kΩ logarithmic potentiometer

C1   10 μF × 16-volt electrolytic capacitor

C2   470 μF × 12-volt electrolytic capacitor

C3, C6   0.1 μF ceramic or polyester capacitors

C4, C5   1,000 μF × 16-volt electrolytic capacitor

PCB, power supply, heatsink for the IC, wires, solder, etc.

The Technology Today

Laser beams are used in several modern appliances. The most well known is probably the laser printer.

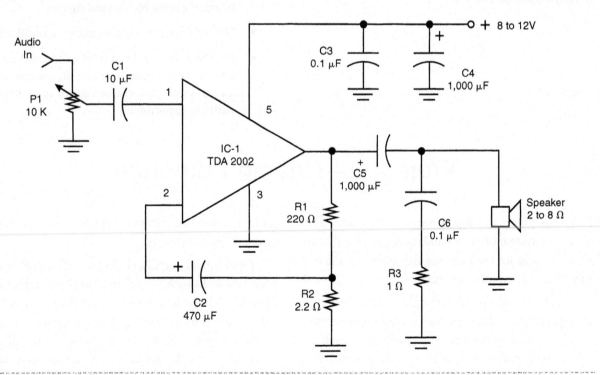

Figure 3.10.26   *High-power amplifier using the TDA2002.*

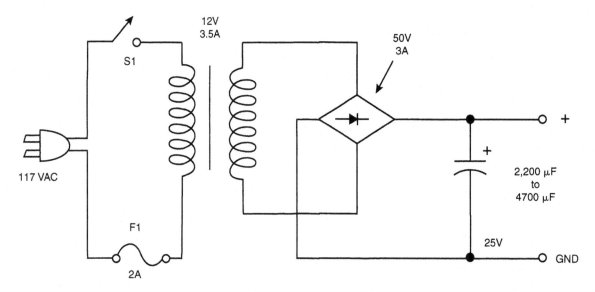

Figure 3.10.27 *Power supply for the amplifier.*

A laser beam is modulated and focused to a photosensitive drum surface. Mirrors are used to alter the direction of the laser beam, while lenses are used to maintain a low divergence at all points along the beam's path.

Another technology that uses mirrors controlled by electronic signals is found in large TV screens. Micromirrors are used to modulate the light beam of each pixel of an image. The mirrors are mounted directly on silicon chips.

Ideas to Explore

The project described here is only a basic one used to perform experiments with laser modulation and control. The evil genius can use his or her imagination to create more advanced devices or higher-performance devices, suggested in the following list:

- Add a circuit to turn on and off the laser beam, creating modulated figures.
- Create Lissajous figures using an oscilloscope.
- Design a light show system so you can modulate the laser beam by using the output of an audio system and create figures that change with the reproduced music.

Project 11—Analog Computer

Today computers are digital machines; that is, they work on a system of zeroes and ones, having only two possible states for their circuits. But this is not the only type of computer that exists.

In the early days of computing, computers were analog machines. They operated according to another principle, which persists in some applications even today. Those analog computers converted numbers into voltages and then performed mathematical operations with those voltages using potentiometers and operational amplifiers.

Operational amplifiers were so efficient at performing tasks such as addition, subtraction, multiplication, division, integration, and differentiation that they are still used today in many applications that use analog signals. Although the operational amplifiers were created to be used for computing, upgraded forms have been created to perform many other applications. Figure 3.11.1 shows the diagram adopted to represent an *operational amplifier* (OA) and some of its configurations.

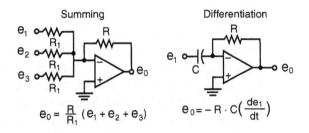

Figure 3.11.1 *An OA and some configurations to perform math operations.*

Converting numbers and sending them as voltages across the circuits, the computer performed mathematical operations and then sent the results to some kind of receiver, as shown in Figure 3.11.2.

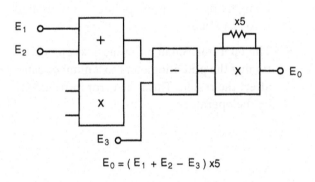

Figure 3.11.2 *Simple structure of an analog computer.*

Analog computers have many deficiencies that digital computers don't present. At each operation the possibility of errors is introduced, which of course would affect the final results. The more operations completed, the greater the chances of having an error affect the results, and therefore the greater the number of errors.

Another problem is speed. Analog computers can't perform a large number of operations simultaneously, making them slow and impossible to use in complex calculations.

Analog computers can be compared to slide rules in their operating principle, as they were soon replaced by modern electronic calculators, which had much greater precision and speed. Figure 3.11.3 shows a slide rule used by engineers in the mid-1900s, before scientific calculators and computers were invented.

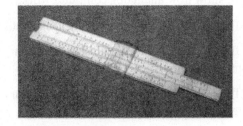

Figure 3.11.3 *A slide rule.*

Purpose

Although analog computers are not widely used today, simple versions such as the ones proposed in this project can be mounted to show how they operate. Because they use mechanical and electronic parts, they are considered to be a mechatronic project, while adding a touch of computer science and even mathematics.

What we will describe in this project is how to build a basic small analog computer that, like a mechatronic slide rule, can perform some mathematical operations such as the following:

- Addition
- Subtraction
- Multiplication
- Division
- Log functions
- Square root functions
- Powers of 2
- Trigonometric functions such as sine, cosine, tangent, and cotangent
- Calculations involving PI (see Project 14)
- Proportional calculus

The circuit is very simple in the basic analog computer but can be upgraded with the use of operational amplifiers and other resources. It can be used for the following objectives:

- To show how an analog computer works
- To perform simple mathematical calculations
- To help students with their homework in mathematics
- To understand how an analog conversion is made
- To be used in fairs, science shows, and in classes

The Project

Figure 3.11.4 shows the diagram of the analog computer that we will build in the basic version. As the reader can see, the basic analog computer has three indicators (attached to the three potentiometers) moving or sliding over a multigraduated scale and an indicator.

Potentiometers 1 and 2 are normally used as data inputs (numbers to be used in a calculation), and potentiometer 3 is adjusted to find the result of the first two potentiometers. When the result of the first two is found, the indicator on the third will show zero.

Figure 3.11.4 *The basic version of the analog computer.*

For instance, if in a multiplication function, you set potentiometer 1 equal to 3 and potentiometer 2 equal to 2, the indicator of potentiometer 3 will show zero (equilibrium) when the potentiometer 3 reaches the position 6 ($3 \times 2 = 6$).

The precision of the results is dependent on many factors:

- **The precision of the potentiometers and the indicator**: Common linear potentiometers are not very accurate devices because they are mechanical. Typically, 5 percent is the normal tolerance rate for these precision components in their scale. If this 5 percent of difference in the results is multiplied by the 3 potentiometers, a great difference in the expected results will occur. Considering the experimental use, this difference will not compromise the project. Additionally, the indicator is not necessarily precise enough to show exactly when the zero point is reached. This can add another small error to the results.

- **The precision of scales**: The scales are not accurate. Like analog multimeters, the actual size of the pointer indicating the numbers can introduce some degree of error. For example, it might appear as if the pointer is pointing to zero or to one.

- **The operator**: Small differences when positioning the potentiometers to a number can affect the results. This is an error introduced by the operator.

How It Works

The operating principle of the analog computer is based on the potentiometer's capability to make analog multiplications. To understand how this works, let's start from the simple circuit shown in Figure 3.11.5.

This circuit can convert the mechanical position of the potentiometer's cursor to an analog voltage. The potentiometer is numbered from 0 to 10, with the

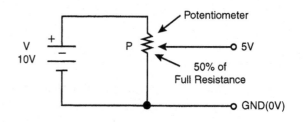

Figure 3.11.5 *Making an analog conversion.*

center position therefore being 5. If the potentiometer is resting in the center or halfway position (5), this means it represents the halfway point or 50 percent of whatever is registering on the scale. This position equals 50 percent of the battery voltage. If 100 percent of the battery voltage is 10 volts, for instance, the corresponding central position (or halfway point) is 5 volts.

Of course, if we intend to work with larger numbers, we may need to make a mental conversion. For instance, if the number of volts to be converted is 47, we can represent it as 4.7 and convert to 4.7 volts. The decimal point position must of course be considered later when the result is found. Figure 3.11.6 shows how to represent other quantities.

Now let's go one step ahead, adding another potentiometer to our circuit, as shown in Figure 3.11.7.

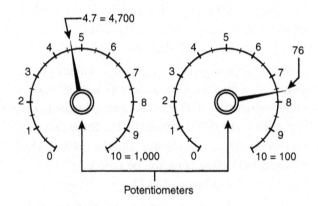

Figure 3.11.6 *Analog conversions using the potentiometer.*

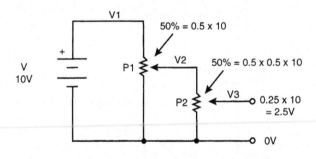

Figure 3.11.7 *Adding a second potentiometer.*

Observing the figure, we can see that the voltage applied to the ends of the second potentiometer is the number selected from the first potentiometer converted to an analog voltage.

So if the second potentiometer is at the central position, or 50 percent of the scale, the voltage in its cursor will be 50 percent of the original 50 percent ($0.5 \times 0.5 = 0.25$) or 25 percent of the battery voltage. In other words, without the decimal point, the circuit is multiplying $5 \times 5 = 25$.

This means that if potentiometers 1 and 2 have scales with numbers from 0 to 10, the voltage in the output of the circuit will be the product of the selected number times 10. In other words, a scale registering 0 to 10 volts will represent numbers from 0 to 100.

But how do you read the results? The answer is shown in Figure 3.11.8, where a third potentiometer and an indicator are added.

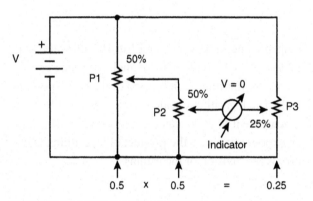

Figure 3.11.8 *Completing the basic analog computer.*

The reader can easily see that if we adjust the third potentiometer (P3) to 0.25 percent on its scale, the voltage in its cursor will be the same as that of the cursor of P2. An indicator placed between those points will indicate zero.

If this potentiometer (P3) is scaled from 0 to 100, the position of its cursor will be the result of the multiplication of the numbers from P1 and P2. This is valid for all numbers, or all points on the scale from P1 and P2, as shown in Figure 3.11.9.

The division of a number can be made with the inverse procedure: The number to be divided is put in P3, the divisor in P2, and the result will appear in P1. When P1 is moved, the circuit detects when the position reaches the result.

The scales, or measurements on the scales, can also be calibrated with other units such as sines, cosines,

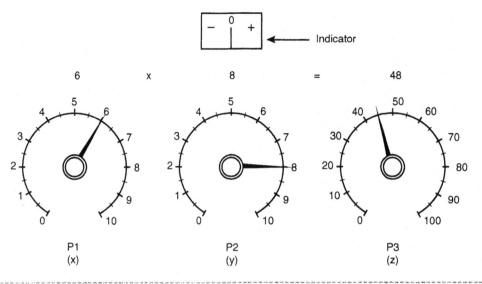

Figure 3.11.9 *Any multiplication can be made using this simple circuit.*

tangents, and logs. Adjusting P1 to 10 (full scale) and P2 to any angle in the scale of angles, the sine or cosine for this angle can be read in the corresponding scale. The logarithmic scale is very important because we can use it to make calculations using logs, because an exponentiation is the product of logarithms, and a root is the division of logarithms.

Components Are Critical

An important problem to be considered in this project is that a potentiometer has finite values of resistances. This means that the circuits can be loaded, thereby affecting the final results. Figure 3.11.10 shows what happens when one potentiometer reports to another potentiometer that cannot accept the message.

In the figure, a 1,000 Ω potentiometer is coupled to another 1,000 Ω potentiometer. However, if P2 is adjusted to 50 percent of the scale (500 Ω), the voltage in A (in Figure 3.11.10) is not 5 volts but less.

What happens is that an equivalent circuit will be formed now by a 500 Ω resistor in series with two other resistors: a 500 Ω resistor formed by the lower half of P1 and 1,000 Ω formed by P2. These two resistors represent a 333 Ω resistor, and the voltage, which we expected to be 5 volts, will be reduced to 3.33 volts, as shown in Figure 3.11.11.

The simplest solution to avoiding this problem is to make P2 as large as possible. If P2 is a 100 kΩ resistor instead of a 1 kΩ, the voltage in P2 will be

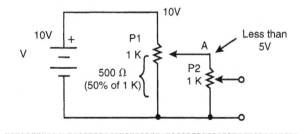

Figure 3.11.10 *P2 is loading P1, affecting the results.*

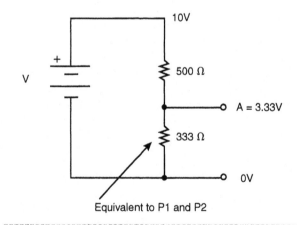

Figure 3.11.11 *The voltage is affected by the second potentiometer's resistance.*

4.95 volts, a value almost equal to the expected 5 volts we had hoped for at that point.

Another important point to consider is that the circuit is not dependent on the power supply's voltage. This is because the voltage is the same in P1, P2, and P3. So any change in one branch of the circuit results in a corresponding similar change in the other branches.

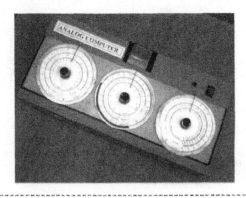

Figure 3.11.13   *Components placed inside a box.*

How to Build the Analog Computer

Figure 3.11.12 shows the basic version of the analog computer. The few components used in the project are placed inside a wooden box, as shown in Figure 3.11.13. The author has built a prototype housed in a 40 × 18 × 8 cendimeter box.

The power supply is formed by four AA cells. P1 and P3 are wire-wound linear potentiometers. P2 is a common carbon linear potentiometer. The reader must take care with the angle of the cursor. All must be set at 270 degrees to the potentiometers to match the scales suggested in the text.

The indicator is a 50-0-50 µA, with the zero at the center of the scale. This kind of indicator is found in some stereo audio amplifiers to indicate balance.

If the reader can't find this indicator, he or she can use any 0 to 200 µA common microamperimeter and add a diode bridge so it will allow currents in both directions. This circuit is shown in Figure 3.11.14.

The diodes must be of the germanium type because the silicon type begins to conduct with as little as 0.7 volts. Germanium diodes conduct with 0.2 volts without causing a large sensitivity loss in the indicator.

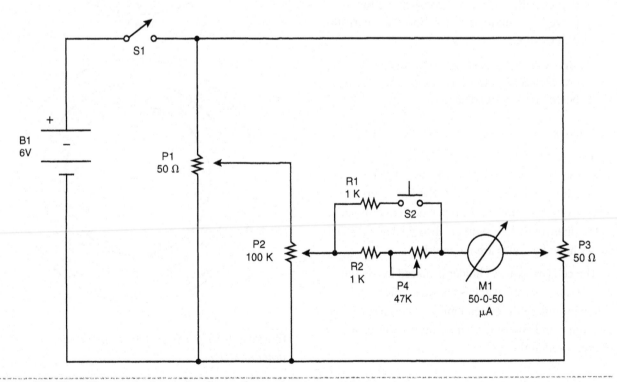

Figure 3.11.12   *Electronic circuit for the analog computer.*

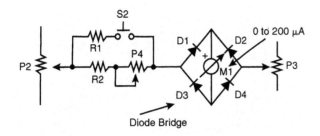

Figure 3.11.14 *Diode bridge allowing the use of common indicators.*

The Scales

The analog computer has the ability to solve problems involving subtraction, addition, multiplication, division, roots, exponentials, logarithmic equations, trigonometric equations, and even combined functions. In the basic version, three dials (X, Y, and Z) are used. Each dial has three scales, or rings of measurement:

- **Linear scales a1, a2, and a3**—These are used in operations such as subtraction, addition, multiplication, division, and to extract square roots.

- **Logarithmic scales b1, b2, and b3**—These are used to make divisions and multiplications by the logarithmic method, to calculate the power of a number, and to find the logarithm of a number.

- **Sine and cosine scales c1, c2, and c3**—These are used to find the sine and cosine of angles between 0 and 90 degrees.

The scales are made as shown in Figure 3.11.15.

You can photocopy these models and glue them to CDs. The CDs are then glued to knobs, such as the ones that fit on potentiometers. The indicators are made from thin wires or even plastic filaments such as the ones found in a synthetic broom. Figure 3.11.16 shows how the scales are mounted.

The cell (or battery) holder is placed inside the box, and the on/off switch is placed on the panel. When mounting the potentiometers, be sure that the beginning and end of the cursor movement falls over the zero and at the end of each scale.

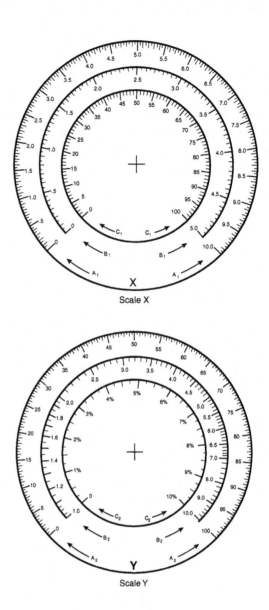

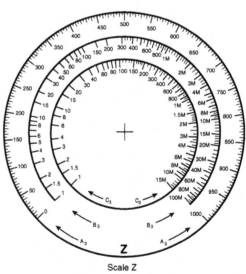

Figure 3.11.15 *The scales for the three potentiometers.*

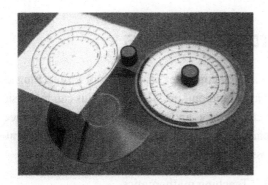

Figure 3.11.16 *Transparent plastic pieces are used with the potentiometers. They are glued to the plastic knobs.*

S1 is used to turn the power supply on and off. S2 is used to increase the sensitivity to a point that will produce the expected result. When S2 is closed, the sensitivity of the circuit increases, making it easier to find the correct result of the mathematical operation made by the computer.

Other Scales

The evil genius may want to make a more powerful analog computer by adding different scales. The following are some examples of scales that can be added:

- Scales for logarithms in the base "e"
- Scales for square and cubic roots
- Scales to convert numbers from the decimal system to other bases or systems
- Scales representing hyperbolic functions

Parts List—Analog Computer

P1, P3	50 Ω linear wire-wound potentiometer (270°)
P2	100 kΩ linear carbon potentiometer (270°)
P4	47 kΩ timer potentiometer
R1, R2	1 kΩ × 1/8-watt resistor (brown, black, red)
S1	SPST switch (on/off)
S2	Pushbutton
B1	6 volts of power (4 AA cells with holder)
M1	50-0-50 µA micrometer with 0 at the center of the scale (see text)

Box, terminal strip, knobs and rules for the potentiometers, scales, wires, solder, etc.

Testing and Using

Place the cells, or batteries, in the cell holder and turn on the S1 switch. P2 must be in the center of the scale. Moving potentiometers P1 or P3 any amount will change the indicator's position.

First adjust P4 so that no indications exist outside the dimensions of the scale. This trimmer potentiometer limits the current across the instrument.

Now you can try simple operations such as a multiplication. Adjust potentiometers P1 and P2 for the numbers to be multiplied. Then adjust P3 to find the result. The indicator on P3 will show zero when the result of P1 and P2 has been found.

Other Operations

- **Division** Mark the number to be divided in the scale a3 of P3 and the divisor in scale a2 of P2. Adjust P1 to read as zero in the indicator. Read the result in the scale a1 of P1.

- **N-root** Start with the following:

  n root of A = b

  Then calculate the *n*-root of A by putting the number A in the log scale of P1 (b1). Fix the inverse of *n* in the linear scale of P2 (a2). The result is read in the log scale (b3) of P3 when the indicator shows zero.

- **Powers of N** Start with the following:

  A powered to b = C

Therefore, fix the base in the log scale of P1 (b1) and the power *b* in the linear scale of P2 (a2). Adjust P3 to read zero. The result is found in the log scale of P3 (b3).

Other operations can follow the same basic rules, many of them found in college mathematics books.

Exploring the Project

Many changes can be made in this basic analog computer project. The evil genius can start from this circuit and upgrade it using modern components such as operational amplifiers. The following are additional suggestions for altering and upgrading the project.

Summing Circuit

Figure 3.11.17 shows how a summing circuit can be added using two or three potentiometers wired in series. The voltage in A is the sum of the analog voltages set in potentiometers P1, P2, and P3.

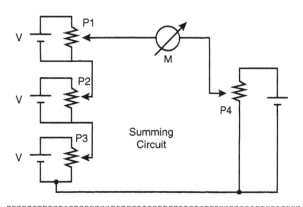

Figure 3.11.17 *Summing circuit using potentiometers.*

Cross Themes

The analog computer is a great tool for teaching and learning mathematics. The evil genius can make simple calculations, and the teacher can use the project in classes to show how slide rules and abaci operate. Crossover in the following areas is suggested:

- Teaching mathematics
- Knowing how potentiometers can be used to make calculations
- Understanding how we can work with analog quantities to make calculations
- Studying how logarithms can make many types of calculations easier

Additional Circuits and Ideas

The basic version is the simplest; we had no electronic active parts in the circuit, and only passive components as resistors and potentiometers. Electronics can be used, and with sensitive circuits the computer becomes more precise and easier to use. The evil genius may want to try the following versions.

Using a Sound Indicator I

Instead of a visual indicator (instrument), you can use an auditory indicator (piezoelectric transducer). The basic difference in the electronic circuitry is that the signal source will be an audio oscillator and the indicator will be a transducer. The basic circuit for this version is shown in Figure 3.11.18.

When the calculator completes an operation, the sound in the transducer disappears. The circuit can be mounted on a small *printed circuit board* (PCB), as is shown in Figure 3.11.19. Take care with the position of the IC. The tone is adjusted by P1.

The oscillator is made from one of the four gates of a 4093 IC and pulls very little current, therefore

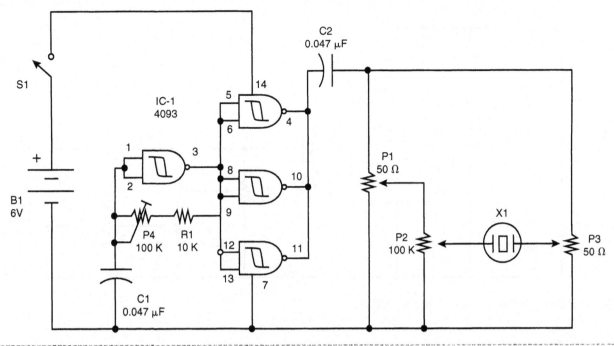

Figure 3.11.18 *A simple audio oscillator is used as the signal source for the computer.*

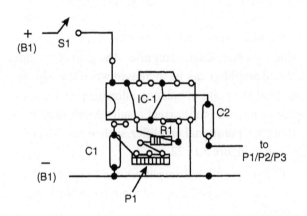

Figure 3.11.19 *The circuit is mounted on a PCB.*

extending the life of the battery. The three other gates are used as digital amplifiers. The transducer must be a piezoelectric high-impedance type. Low-impedance transducers won't work in this circuit.

Parts List—Using a Sound Indicator I

IC1	4093 *complementary metal oxide semiconductor* (CMOS) IC
R1	10 kΩ × 1/8-watt resistor (brown, black, orange)
P1, P3	50 Ω linear wire-wound potentiometers (270°)
P2	100 kΩ linear carbon potentiometer (270°)
P4	100 kΩ trimmer potentiometer
C1, C2	0.047 μF ceramic or polyester capacitor
S1	SPST switch
X1	Piezoelectric transducer
B1	6 volts of power (4 AA cells with holder)

PCB, wires, box, knobs, solder, etc.

Using a Sound Indicator II

Another circuit using sounds to detect the null point is shown in Figure 3.11.20. This circuit is based on the well-known 555 IC and is configured as a stable multivibrator, generating a square signal, with the frequency adjusted by P4. The circuit can be mounted using a small PCB, as shown in Figure 3.11.21.

The operating principle is the same as in the previous project. The null detector is a piezoelectric transducer or a high-impedance transducer.

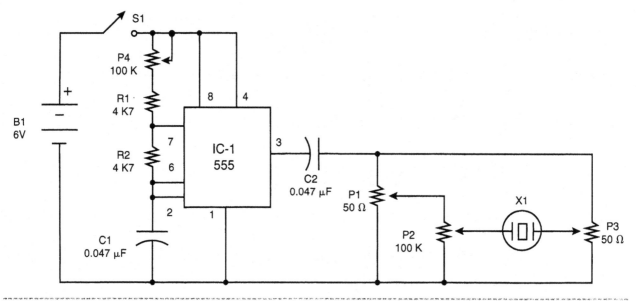

Figure 3.11.20 *Sound circuit using the 555 IC.*

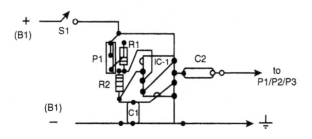

Figure 3.11.21 *Printed circuit board for the sound circuit.*

Parts List—Using a Sound Indicator II

IC-1	555 IC timer
P1, P3	50 Ω linear wire-wound potentiometer (270°)
P2	100 kΩ linear carbon potentiometer (270°)
P4	100 kΩ trimmer potentiometer
R1, R2	4k7 Ω × 1/8-watt resistor (yellow, violet, red)
C1, C2	0.047 μF ceramic or polyester capacitor
X1	Piezoelectric transducer
B1	6 volts of power (4 AA cells with holder)

PCB, wires, knobs for the potentiometers, solder, box, etc.

Adding an Operational Amplifier—LEDs Indicator

Sensitivity can be added to the balance circuit by adding an operational amplifier. The gain of the operational amplifier can be increased when the adjusts are near the point of balance, allowing much more precision to find the correct results. A circuit using a common operational amplifier, such as the 741, is shown in Figure 3.11.22.

Other operational amplifiers can be used in the same configuration, many of them requiring lower voltages.

When the circuit is in use, the low-gain position is recommended. When the position corresponding to the mathematical result is almost found, the switch S2 is put in the position for high gain. This circuit can be used to detect two LEDs. When the balance is found, both LEDs will turn off.

Parts List—LED Indicator Operational Amplifier

IC-1	741 IC operational amplifier
LED1, LED2	Common LEDs (any color)
R1	1 kΩ × 1/8-watt resistor (brown, black, red)

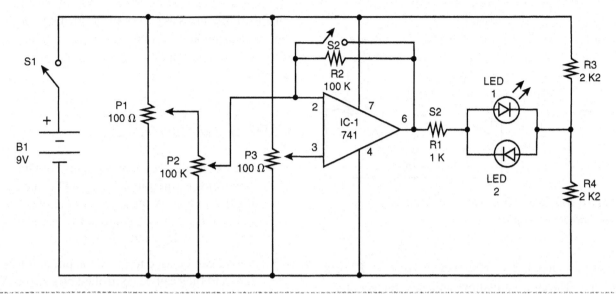

Figure 3.11.22 *Circuit using an operational amplifier.*

R2	100 kΩ × 1/8-watt resistor (brown, black, yellow)
R3, R4	2.2 kΩ × 1/8-watt resistor (red, red, red)
S1, S2	SPST switch
B1	9 volts of power (6 AA cells)
P1, P3	100 Ω linear wire-wound potentiometer (270°)
P2	100 kΩ carbon linear potentiometer (270°)

PCB, cell holder or battery connector, wires, solder, etc.

An Advanced Circuit with Four Potentiometers

More complex calculations, including equations, can be solved with a four-potentiometer version of the analog computer. As shown in Figure 3.11.23, the circuit can solve equations of the following type:

$$A/B = C/D$$

Using this circuit to solve an equation, any of the four potentiometers can be the x. For instance, in the equation $3x = 2 \times 4$, the potentiometers are adjusted in the following order:

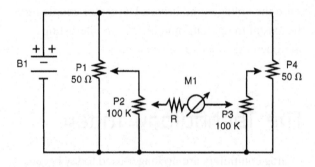

Figure 3.11.23 *The circuit can perform calculations of the type A/B = C/D.*

P1 = 3

P3 = 2

P4 = 4

And you will move P2 (x) to balance the circuit and solve the equation.

An Automatic Mechatronic Analog Computer

Gearboxes and a comparator can be used in an advanced project of an analog computer, as Figure 3.11.24 shows.

This circuit senses the voltage resulting from the operations of P1 and P2, and compares it with the voltage in the cursor of P3. If the voltages are not the

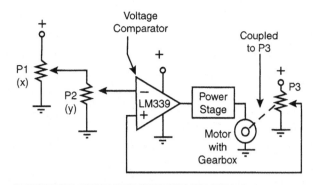

Figure 3.11.24 *Using servos to build a mechatronic computer.*

same, the circuit acts and moves the cursor of P3 with a servo. The cursor will move until the voltages are the same.

A comparator (LM339, for instance) is used to sense the voltages and drive a DC motor coupled to the axis of the potentiometer. It is an interesting approach for a mecatronic project.

The Technology Today

Analog computers are no longer used today because digital computers are much more precise and powerful. However, analog computers are studied in many engineering courses as part of the curriculum on operational amplifiers. Analog computers help to know how these devices operate.

Ideas to Explore

- The evil genius can imagine many configurations for the potentiometers, possibly increasing the number of them and adding switches to make the changes. A powerful analog computer can be designed this way.

- Servos can be used to move the potentiometers, and they can be controlled by sensors. A computer can perform calculations regarding physical quantities measured by the sensors.

- Digital potentiometers can be used in a project where the values of the quantities to be used in the calculations can be set automatically by a computer or set digitally.

- The Internet is a great resource for information about analog computers. If you are interested in this subject, you will find many sites with several practical approaches to be used in these projects.

Project 12—Touch-Controlled Motor

Many interesting experiments can be performed using this simple touch control for small DC motors. Start your robot or open a window with a simple touch of your fingers. Scientific experiments can also be done where a person or animal starts a motor using this control, and simple games to test your friends' skill can be built using a touch-controlled motor.

Motors requiring currents up to 3 amps and with voltages of between 3 and 15 volts can be controlled by this circuit. That means that almost all small DC motors can be controlled by this circuit.

Objectives

- Build a circuit that can control motors with the touch of your fingers.

- Show how skin resistance is enough to trigger a circuit.

- Create automatic routines using a touch-controlled motor.

- Teach the concept of a closed circuit to explain how the project works.

How It Works

The basic circuit uses a monostable version of the 555 *integrated circuit* (IC) to trigger a motor during a programmed timer interval. When touching a sensor, the current flowing across the skin is enough to trigger the circuit and a motor activating and maintaining the operation mode during a timed interval determined by an RC network. The timed network can be programmed for intervals ranging from some milliseconds to several minutes.

As the aim is to increase the sensitivity of the circuit, a transistor is added to supply the trigger voltage to the IC from the touch sensor. In this way, currents as low as a few microamperes can trigger the circuit to power on the motor, as shown in Figure 3.12.1. The current necessary to trigger the circuit comes from a sensor that the controller must touch.

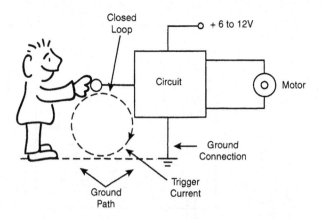

Figure 3.12.2 *The current must have a closed loop to trigger the circuit.*

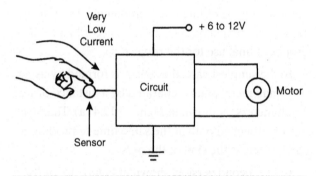

Figure 3.12.1 *A current of a few microamperes is enough to trigger the circuit powering the motor.*

The skin resistance of a person depends on several factors, including humidity, thickness, and amount of salt (due to transpiration). Other factors that determine the viability of the current is the path of the current, as shown by Figure 3.12.2.

The current needs a closed path to act on the circuit. This means that eliminating leaks of current through the shoes or through a connection to the ground is important to increasing the sensitivity of the circuit.

When the person touches the sensor, the current flowing across the skin, body, and ground determines the operation of the circuit. In general, because this current does not involve more than a few microamperes, it isn't enough to cause any shock sensation or danger to the person. Only a brief touch on the sensor is enough to trigger the circuit that activates and maintains the on state at the programmed time interval.

Although no shock is caused when touching this sensor, it is still important to consider the possibility of shock when working with electronics. To be absolutely secure, you must take the following caution.

> **Caution!** Do not power the circuit from transformerless power supplies or supplies directly connected to the AC power line! They are not isolated from the danger of causing a severe shock hazard.

Once triggered, the circuit remains on due to the configuration of the 555 IC in the monostable mode. In this mode, when the input (pin 2) is grounded for an instant, the output of the circuit (pin 3) goes to the high logic level. This means that a voltage near the power supply voltage can be sourced by this pin.

The output remains in the on state through a time interval determined by the capacitor C2, and the value of the resistance is adjusted by P1 and R4. Considering R (the value of the adjusted resistance), the time interval in which the output remains on can be calculated by the following formula:

$$t = 1.1 \times R \times C$$

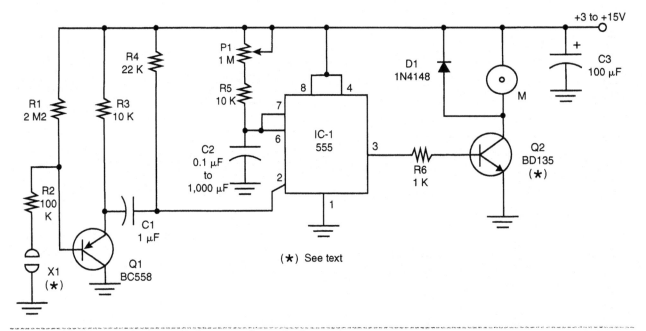

Figure 3.12.3 *Touch-controlled motor schematic diagram.*

where:

t = the time interval in seconds

R = the resistance in ohms (R4 + P1)

C = the capacitance in farads (C2)

The output of the 555 IC can source about 200 mA, but it is not recommended to couple it directly to high-power loads such as a motor, due to the possibility of instabilities.

Therefore, a driver stage, using a transistor (Q2), is added. With this driver stage, small motors with current drains of up to 500 mA can be driven. If you want more current, use a TIP31 or even a Darlington power transistor. The transistor must be attached to a heatsink.

P1 can be a trimmer potentiometer, an external control, or a log or linear potentiometer. The touch sensor can be built in two methods according to the proposed final use for the project.

In the simplest case, the sensor is formed by using two small metal plates separated by a distance of 1 or 2 millimeters, as shown in Figure 3.12.4 (a). The plates must be touched both at the same time to produce a current across the skin of the person.

In the second version, shown in Figure 3.12.4 (b), only one sensor is used, because the current flows across the body of the person. In this case, the circuit must have a ground connection to offer a path for the current. The ground connection can be made using any metal object contacting the ground, such as a metal window or door.

How to Build

The complete schematic diagram for the touch-controlled motor in the basic version is shown in Figure 3.12.3. The circuit can be mounted on a *printed circuit board* (PCB) or, for experimental purposes, on a solderless board. The circuit can be powered by cells or a power unit connected to the AC power line, but be sure to reread the previous Caution!

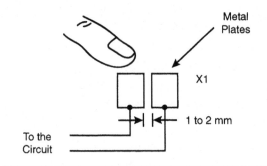

Figure 3.12.4 *Touch sensors.*

Figure 3.12.5 shows a simple power supply suitable to control motors and rated to 6 or 12 volts and with currents up to 1 amp.

The transformer has a primary coil rated to 117 VAC or, according to the AC power line of your location, a secondary coil rated to 7.5 volts (6 volts of output) or 12 volts (output), and currents up to 1 amps, according to the motor.

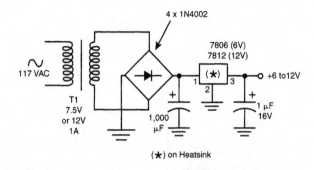

Figure 3.12.5 *A simple regulated power supply for the circuit.*

Testing and Using

Power on the circuit and adjust the time control P1 to the minimum resistance (lower time interval). If the motor turned on when you powered the circuit, it will stop in few seconds.

Then touch the sensor. The motor will start and run for a few seconds, according to the value of C2. If you are using a large capacitor, the time it runs before stopping can continue for more than a few seconds, so please wait. If the circuit operates as expected, you can install it in the application you have in mind.

Parts List—Basic Touch-Controlled Motor

IC-1	555 IC timer
Q1	BC558 or equivalent (any *positive-negative-positive* [PNP] general-purpose transistor)
Q2	BD135 medium-power PNP transistor
D1	1N4148 or 1N914 general-purpose silicon diode
R1	2.2 MΩ × 1/8-watt resistor (red, red, green)
R2	100 kΩ × 1/8-watt resistor (brown, black, yellow)
R3, R5	10 kΩ × 1/8-watt resistor (brown, black, orange)
R4	22 kΩ × 1/8-watt resistor (red, red, orange)
R6	1 kΩ × 1/8-watt resistor (brown, black, red)
P1	1 MΩ trimmer potentiometer or common potentiometer
C1	1 μF electrolytic or other capacitor
C2	0.1 to 1,000 μF electrolytic or other capacitor
C3	100 μF × 16-volt electrolytic capacitor
M	Any DC motor with 3 to 15 volts (up to 500 mA)
X1	Touch sensors (see text)

PCB or solderless board, heatsink for Q2, wires, power supply, etc.

Exploring the Circuit

You can use a touch-controlled motor to design many interesting automatic devices for your home or for your robots and mechatronic devices.

Automatic Window

A gearbox and the touch-controlled motor can be used to open and close a window or screen, as shown in Figure 3.12.6.

A switch must be added at the end of the window routine to reset the circuit or to invert the direction of the motor.

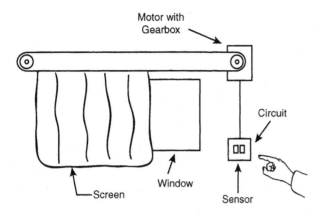

Figure 3.12.6 *Opening and closing a curtain with the circuit.*

Touched-Controlled Robot or Car

Another interesting application for the touch control is to start a small robot that will run for a time interval adjusted by P1. A race car such as the one described in Project 1 can be used. After installing the sensor in such a car, a touch to the sensor will turn on the motor for a specific time interval. In a competition, adjust all cars for the same time interval. The winner might be the one who can cover the longest distance, as shown by Figure 3.12.7. Other toys can be controlled by the touch using the same circuit.

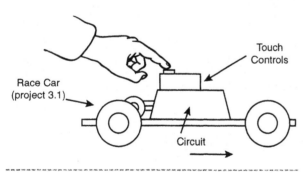

Figure 3.12.7 *A robot or race car controlled by touch.*

Timed Fan

A small fan using a DC motor can be activated using the touch control. It will remain on for the duration of the time interval set by P1. Using a 1,000 μF capacitor and setting P1 to the maximum resistance, each touch will turn on the fan for a time interval of about 15 minutes.

Automatic Hand-Drying Machine

The basic idea for this automatic device is to create a fan, activated by touch, that blows hot air in the direction of your wet hand, as shown in Figure 3.12.8.

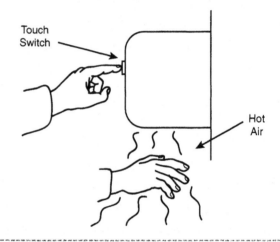

Figure 3.12.8 *Hand-drying machine.*

Cross Themes

Other experiments involving a motor controlled by a sensor can be performed to illustrate many subjects in the elementary and high school curriculum. The following is a list of suggestions.

Traps

Small animals can trigger the circuit when touching the sensor and then closing the trap, as shown by Figure 3.12.9.

The circuit must be adjusted to run the motor during the time interval necessary to close the trap. An alternative is to include a switch at the end of the course of the trap that will turn off the motor.

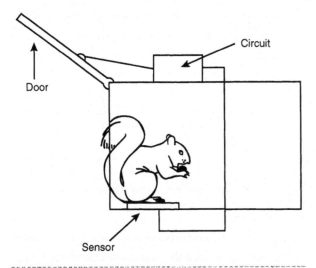

Figure 3.12.9 *A trap using the touch-controlled motor.*

Mixer

An experimental mixer that can be used for mixing lab substances, juices, or other applications can be mounted as shown in Figure 3.12.10.

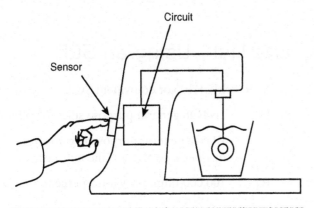

Figure 3.12.10 *A mixer controlled by touch.*

Newton's Color Wheel

One of the most traditional experiments made when studying optics is Newton's color wheel. Through this experiment we can show that the result of the combination of the fundamental colors is white. Figure 3.12.11 shows how to use the touch-controlled motor to build a Newton's color wheel.

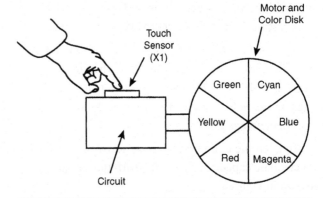

Figure 3.12.11 *Touch-controlled Newton's color wheel.*

It is important to follow exactly the recommended colors for the wheel. If different combinations of colors are made, the result will not be white!

Mechatronic Race Car

The mechatronic race car described in Project 1 can be controlled by touch using this circuit. Use this circuit and program the timer (by P1) to allow enough time for the car to complete the programmed distance for the competition.

Additional Circuits

Many other electronic configurations can be used with your touch sensor. The following is a short list of some of the configurations you might want to try.

Using an Silicon Controlled Rectifier (SCR)

This version of the touch motor control uses a *silicon controlled rectifier* (SCR) as a sensitive switch placed in series with the DC motor to be controlled. SCRs such as the TIC106 proposed in this circuit are very sensitive. They can be triggered by currents as low as a few hundred microamperes.

Another characteristic to be considered in this circuit is the self-latched operation of an SCR. That means that, when triggered, an SCR will remain in the on state even if the trigger voltage disappears.

Some small DC motors use brushes to switch the coils, thus inverting the current flow when in operation. These brushes turn on and off at a high speed, causing the SCR to turn off as long as the sensor is not touched again.

To latch the circuit on, you can add a 10 to 100 μF electrolytic capacitor in parallel with the motor, as shown in Figure 3.12.12.

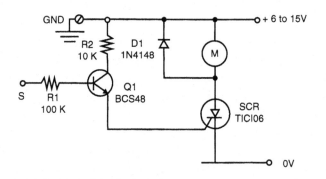

Figure 3.12.13 *Touch control using an SCR.*

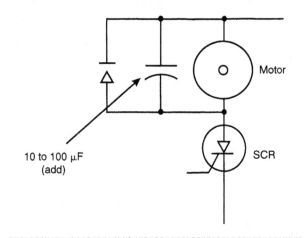

Figure 3.12.12 *Latching the circuit.*

Other important facts to be considered in this version include the following. The *ground connection* (GND), which increases the sensitivity as the current flows in a closed loop, passes by the body of the person who touches the sensor. Without the ground connection, the trigger current (the current produced by the object touching the sensor) is reduced. As sensitivity is reduced, the object touching the sensor should be large enough to produce the amount of current needed to trigger the circuit.

Also consider the voltage fall across an SCR in the on state. When the SCR is on, a voltage fall of about 2 volts occurs. This means that the DC motor receives 2 volts less from the power supply. In some applications, this fall should be compensated for by adding 2 volts to the power supply output. The schematic diagram of the touch-controlled motor using an SCR is shown in Figure 3.12.13.

Any DC motor rated at 6 to 12 volts and with a current drain of up to 1 amp can be used. If motors rated at more than 6 volts are used, change the power to one that can source the voltage needed to power the motor. The circuit can be housed in a small box, and the sensor can be placed as far away from the circuit as necessary. The sensor is a 5- × 5-centimeter metal plate, and the wire to connect it to the circuit can be up to 10 feet long.

Use cells to power the circuit. However, if you choose a different type of power supply, avoid transformerless power supplies because they are not isolated from the AC power line and can cause severe shock hazards.

Parts List—Using an SCR

SCR   MCR106 or equivalent SCR

D1   1N4148 general-purpose silicon diode

Q1   BC548 general-purpose *negative-positive-negative* (NPN) silicon transistor

R1   100,000-ohm, 1/4-watt, 5 percent resistor

R2   10,000-ohm, 1/4-watt, 5 percent resistor

S   Sensor (see text)

GND   Ground connection (see text)

Terminal strip, power supply, plastic box, wires, solder, etc.

Using the 4093

CMOS ICs have very high input impedance. This great sensitivity is ideal for applications in a touch switch.

Many common CMOS ICs can be used to implement a touch switch, and the 4093 in particular is a perfect candidate for this task. Figure 3.12.14 shows how one gate of the 4093 can be used in a touch switch to drive a DC motor through a drive stage.

In this circuit, the resistor R1 determines the input sensitivity. The values can range from 1 MΩ (low sensitivity) to 47 MΩ (high sensitivity). As in the other circuits, the power supply must not be transformerless. The three other gates of the 4093 are used to drive the power output stage, formed by an NPN transistor.

Parts List—Using the 4093 IC

IC-1   4093 CMOS IC

D1   1N4148 general-purpose silicon diode

Q1   BD135 or TIP31 NPN silicon power transistor

R1   22 MΩ × 1/8-watt resistor (red, red, blue)

R2   1 kΩ × 1/8-watt resistor (brown, black, red)

X1   Touch sensor (as in the basic project)

M   6- to 12-volt DC motor, up to 500 mA (BD135) or up to 2 amps (TIP31)

PCB or solderless board, heatsink for the transistor, power supply, wires, etc.

Using the 4013

The 4013 CMOS IC is formed by two independent D-type flip-flops powered by 5- to 15-volt sources. In the circuit shown in Figure 3.12.15, the 4013 is used with a 555 monostable IC to drive a motor using sensors or a pushbutton.

Touching the sensor the first time activates the motor and keeps it on until the sensor is touched again. When touched again, the motor turns off and

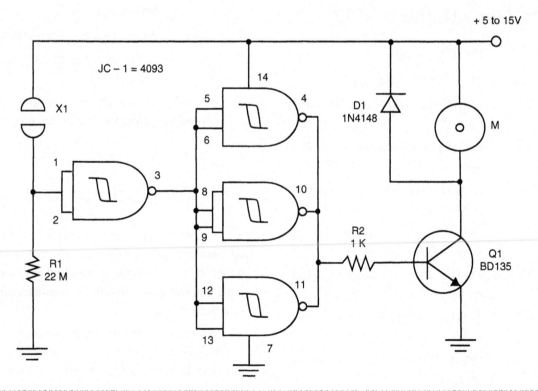

Figure 3.12.14   *Using the 4093 IC.*

Section Three   The Projects

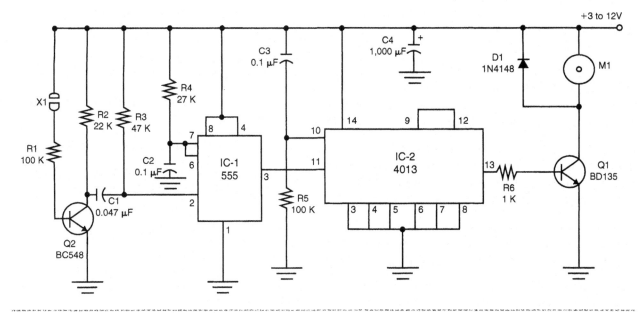

Figure 3.12.15 *The 4013 allows a bistable operation for a motor.*

remains in the off state until the sensor is touched again.

Using the BD135 motor, up to 500 mA can be controlled with this circuit. It is important to decouple the motor to avoid unstable states of the circuit due to the fast switch of the brushes.

Parts List—Using a 4013

- IC-1   555 IC timer
- IC2   4013 CMOS IC flip-flop
- Q1   BD135 silicon NPN medium-power transistor
- Q2   BC548 general-purpose NPN silicon transistor
- D1   1N4148 general-purpose silicon diode
- R1, R5   100 kΩ × 1/8-watt resistor (brown, black, yellow)
- R2   22 kΩ × 1/8-watt resistor (red, red, orange)
- R3   47 kΩ × 1/8-watt resistor (yellow, violet, orange)
- R4   27 kΩ × 1/8-watt resistor (red, violet, orange)
- R6   1 kΩ × 1/8-watt resistor (brown, black, red)
- C1   0.47 µF ceramic or polyester capacitor
- C2, C3   0.1 µF ceramic or polyester capacitor
- C4   1,000 µF × 15-volt electrolytic capacitor
- X1   Sensor (see basic project)
- M1   Motor with 3 to 12 volts, up to 500 mA

PCB or solderless board, heatsink for the transistor, power supply, wires, etc.

The Technology Today

Many types of sensors can be used to start a motor or an automatic process. For example, in airports, shopping centers, and buildings, you find automatic systems that open and close doors commanded by signals from sensors.

They often use infrared sensors, touch sensors, or weight sensors. Elevators are examples of automatic devices where touch sensors are used; touch the button and the message is sent for the elevator to come to your floor. Many appliances in your home also use touch switches. All these devices operate based on the same principles applied in this project.

Exploring the Project

- Use this circuit to show how a small amount of current flowing through your body can control a DC motor that requires a large amount of current.

- Attach an object to the motor as a game, as shown in Figure 3.12.16. Place several sensors in a plastic base and wire some of them to the circuit input. Hide the wires from the view of the players. Imagine that each wired plate is a mine. The player should touch the sensors but avoid the explosion that will occur if a wired

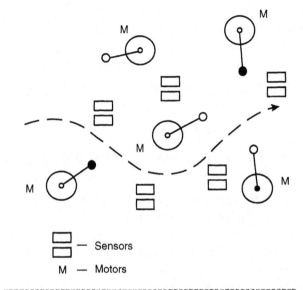

Figure 3.12.16 *Mine field construction.*

plate is touched. The player tries to pass the "mine field" without triggering the motor.

Project 13—Mechatronic Elevator

This project describes a mechatronic elevator built with common mechanical parts and controlled by a simple electronic circuit. It is an ideal project to be adopted in high schools to teach this idea or to test the skills of the mechatronics evil genius.

This kind of project uses both mechanical and electronic parts and has a special visual effect. You can use the elevator as a toy to decorate a miniature building or in science fairs.

The basic project described here has a minimum of resources, meaning that the reader with a good imagination can add other resources or even intelligence by controlling it with a computer. Many options will be given.

The elevator is not powerful because it uses a small DC motor powered by a DC power supply or even common cells. However, depending on the gearbox or the torque system, it can be used to lift weights up to 100 grams.

This project is used as part of the mechatronic course for high school students at Colegio Mater Amabilis in Brazil. The pupils have to put their imagination to work to create the best elevator using gear-

boxes and reduction systems made with simple parts such as diskettes, rubber bands, CD boxes, and other common items (see Figure 3.13.1).

It is important to note that by using the same principles applied in this project other automatic devices such as miniature mobile bridges and industrial machines, doors, and windows can be created.

Figure 3.13.1 *The elevator uses simple parts.*

Objectives

- Build a simple elevator using common parts.
- Add an automatic system to make the elevator stop at different floors.
- Learn how gearboxes work.
- Work with formulas to calculate differential pulleys.
- Organize a competition to see who can build the most powerful elevator.

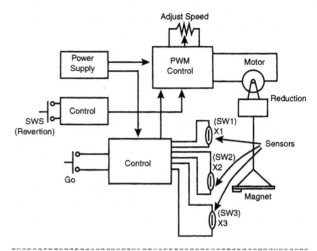

Figure 13.3.2 *The block diagram for the control circuit used in the elevator.*

How It Works

The project of the mechatronic elevator can be divided into two parts: (1) the electronic circuit, which is used to control the motor, and (2) the mechanical part, which includes the gearbox or reduction system and the differential pulley.

Electronic Circuit

To allow the best control of the motor, including the speed, the solution adopted uses a *pulse width control* (PWC). The circuit is the same as the one used in Project 3. The reader can find there a detailed explanation about the operating principle of this kind of control.

The *pulse width modulation* (PWM) control is coupled to a directional control to allow the motor to move the elevator up and down. The directional control and an intelligent block are added to avoid having the elevator push against the top and bottom of the course and to stop at the programmed floors. Figure 3.13.2 shows a block diagram for the electronic circuit.

The sensors are made with reed switches. Attaching a small magnet to the elevator, it will close the reed switches when moving up and down at each floor, as shown in Figure 3.13.3.

The circuit uses a 6-volt DC motor. Therefore, the reader has two alternatives for the power supply. The first is the one shown in Figure 3.13.4 that can deliver

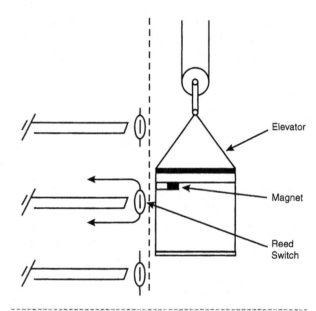

Figure 3.13.3 *Using reed switches as sensors to detect the elevator position.*

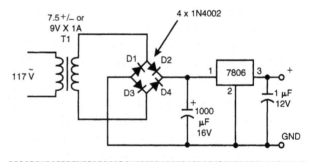

Figure 3.13.4 *Power supply for a 6-volt version of the elevator.*

up to 1 amp to a 6-volt circuit. The second option is to use four C or D cells.

The transformer used in this version has a primary coil rated for 117 VAC, or in accordance with the AC power line, and a secondary coil rated to 7.5 or 9 volts with currents up to 1 amp. The *integrated circuit* (IC) must be mounted on a heatsink, and the C1 capacitor must be rated to 16 volts or more.

The K1 relay is used to stop the elevator at each floor (or level) as determined by the sensors SW1 to SW3. The sensors are reed switches activated by a small magnet attached to the elevator.

Each time the elevator stops, the operator must wait a moment and then press SW4 to start the circuit again and allow the elevator to go on its way.

At the end of the course of the elevator (that is, at the highest and lowest points for the elevator), the reader must add sensors to invert the direction of the motor. But for the simplest project this task can be performed by a switch, as shown in Figure 13.3.2. Magnets, end-of-course switches, H-bridges, and mechanical sensors are other solutions for this task.

The trimmer potentiometer allows the reader to adjust the speed for the best operation of the system according to the power of the motor and the characteristics of the mechanical system.

Mechanical Part

The reader can adopt many mechanical solutions to build this elevator. Common materials such as wood, plastic, or even metal can be used to make the structure of the beilding that holds the elevator.

For the elevator, two solutions should be considered for transferring the power from the motor to the cabin. One is shown in Figure 3.13.5 and uses a reduction box made with wheels and rubber bands. The wheels were made from diskettes and attached to a CD box. It is a very simple solution adopted by many students who build this elevator.

Additional power for increasing and reducing the speed can be achieved using a differential pulley made with plastic or metal and cotton wire. Figure 3.13.6 shows how a differential pulley can be made.

Figure 3.13.6 *A differential pulley in an experimental elevator.*

How to Build

Figure 3.13.7 shows the schematics for the electronic circuit adopted in the elevator's project. The PWM control is the same as the one described in Project 3. The reader can learn more about this part of the circuit by reviewing that project.

The recommended motor is a 6-volt DC motor, rated to enough power to move the elevator. For small motors, such as the ones that can be driven by AA cells, the elevator can handle objects weighing a maximum of 50 grams. Of course, this depends also on the maximum speed and the reduction ratio given by the mechanical parts. By using these small motors, the power can also be delivered by four C or D cells.

The sensors are small reed switches placed in the structure of the building and activated by a small magnet glued to the elevator. It is important to ensure that the elevator does not escape from its trajectory (or elevator shaft) while it is moving, which would change the alignment and disallow the magnet from passing directly in front of the sensors. A system of rails can be used to avoid this alignment problem. The circuit can be mounted on a solderless board or *printed circuit board* (PCB), according to the resources of the reader.

Figure 3.13.5 *A reduction system made with diskette wheels and rubber bands.*

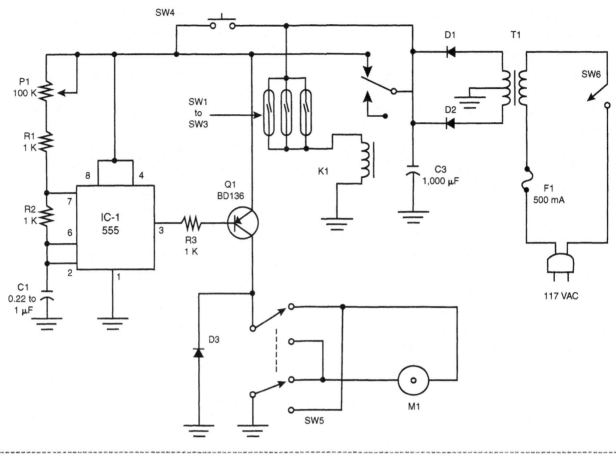

Figure 3.13.7 *The schematic diagram for the electronic circuit.*

Parts List—Elevator with Power Supply Included

CI-1	555 IC
Q1	BD136 *positive-negative-positive* (PNP) medium-power silicon transistor (equivalents such as the TIP32 or TIP42 can be used)
D1, D2, D3	1N4002 silicon rectifier diode
R1, R2, R3	1 kΩ × 1/8-watt resistor (brown, black, red)
P1	100 kΩ potentiometer
C1	0.22 μF to 1 μF ceramic or polyester capacitor
C2	470 μF × 12-volt electrolytic capacitor
C3	1,000 μF × 12-volt electrolytic capacitor
T1	Transformer: primary coil at 117 VAC or according to the power supply line, secondary coil at 6 volts × 500 mA or according to the motor
K1	6 volts × 50 mA SPST relay
SW1, SW2, SW3	*Normally open* (NO) reed switches
SW4	Pushbutton NO
SW5	DPDT switch (see text)
SW6	SPST on/off switch
M1	6-volt DC motor (see text)
F1	500 mA fuse and holder

PCB or solderless board, wires, plastic box for the control, mechanical parts for the building and elevator, small magnet, solder, etc.

Testing and Using

First test the electronic control for the motor away from the mechanical system. Adjust P1 and see if the motor will run gently at all speeds. If you have any problems controlling the speed, change the value of C1. Depending on the motor, lower values can cause problems at high speeds, and higher values can make the motor vibrate at low speeds. To activate the motor run in this test, it is enough to place the magnet near any sensor (SW1 to SW3) or press SW4.

Once you have verified that the motor runs well, attach it to the elevator and test the mechanical system that triggers the sensors to see if the motor has enough power to move the cabin.

Adjust P1 to the ideal speed. Do not allow the elevator to run too fast. Be sure the rubber bands are correctly adjusted to the wheels in the reduction system so they will not slide when rotating.

Once tested, you can use your elevator in demonstrations or science fairs, or you can upgrade it with the suggestions in the next section.

Exploring the Project

Starting with the basic project, some interesting ideas such as the following ones can be explored.

Using a Gearbox

Gearboxes are very efficient methods for increasing the power and adding speed reduction to DC motors. Many types of gearboxes can be found in model stores or in toys. The reader can use a gearbox in the elevator, replacing the mechanical system that was made with rubber bands. Figure 3.13.8 shows a common gearbox that can be used in the elevator.

When using a gearbox, the reader must attempt the reduction ratio given by the mechanical system. The reduction ratio is the factor by which the speed is reduced and the power increased. For instance, a

Figure 3.13.8 *A common gearbox.*

1:100 gearbox raises the torque of the motor by a factor of 100 and reduces the speed by the same amount.

Gearboxes with reduction factors between 50 and 100 are recommended for this elevator. Gear motors or gearboxes including DC motors can also be used.

Adding Other Automatic Routines

Besides automatically stopping at each floor and reversing the motor at the end of course, many other automatic routines can be added to the mechatronic elevator.

The evil genius who is experienced in electronics can design many small circuits or blocks to make the elevator operate in different ways. The following is a short list of suggestions for additions to your elevator:

- Turn on a light in the building when the elevator stops at a particular floor.
- Add call buttons that make the elevator go to the floor where the button was pressed.
- Add a microcontroller programmed to make the elevator accomplish predetermined tasks.
- Add sensors to detect when an object is not perfectly aligned in the elevator.
- Design a digital indicator to show the floor where the elevator is resting or passing.

Controlling the Elevator by Using a Computer

Figure 3.13.9 shows a simple interface that can be used to control the elevator with the help of a computer.

An additional circuit that could be added might be a data acquisition interface plugged into the same port. This data circuit will allow the circuit to send information to the computer about the position of the sensors. A program can be created to control the elevator automatically by using signals sent by the sensor.

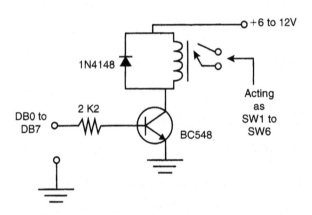

Figure 3.13.9 *An interface to control the elevator through the parallel port of a computer.*

Using a Stepper Motor

Project 14 shows in detail how to use a stepper motor in mechatronic projects. This type of motor can be used with many advantages in applications where precision and fine movements are needed. Another advantage of using a stepper motor is that it is easier to control with a computer.

Cross Themes

Pulleys, differential pulleys, gears, and the work it takes to move an object are all studied in physics. Thus, the elevator project can teach students a great deal about those subjects if used as a crossover project in high-school-level curriculum.

Calculating Pulleys and Gearboxes

Pulleys and gearboxes can be calculated by using the formulas shown in Figure 3.13.10. As the reader can see, the torque and speed are altered according to the ratio of the diameter (or the number of teeth) of the pulleys or gearbox.

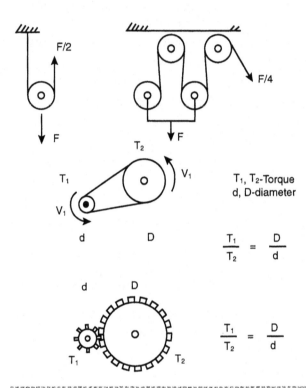

Figure 3.13.10 *Calculating gearboxes and pulleys.*

Calculating Differential Pulleys

Differential pulleys are calculated according to the formula shown in Figure 3.13.11.

Other mechanical machines that present some mechanical advantage can be studied based on the same principles used in the elevator.

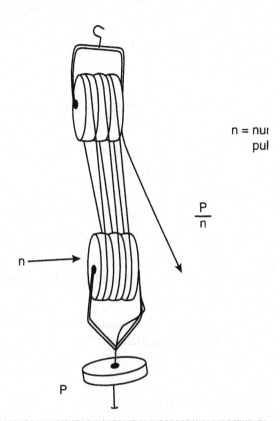

Figure 3.13.11 *Calculating differential pulleys.*

The Elevator Challenge

Who can build the best elevator? A competition can be designed to determine the elevator with the best performance. Rules should be determined regarding the specifics of the elevator. The elevator that can lift the most weight or the one with the greatest number of added functions could be the winner.

Additional Circuits and Ideas

Many circuits or functional blocks can be added to create a fantastic elevator. A few configurations using common parts are suggested in the next section.

Using a Linear Control

The reader can replace the PWM control with the electronic potentiometer described in Project 7. This circuit is not as good as the PWM control because the torque and speed can change according to the power applied to the motor. The torque and speed in this circuit can also change with the weight to be lifted in the elevator, so adjusting the circuit requires more care.

Adding a Timer

Figure 3.13.12 shows a simple timer that can be used to stop the elevator at each floor at a particular time.

When the input (pin 2) of the 555 timer is put to the ground level by any sensor, the output goes to the high logic level and the relay turns on, cutting the power for the motor.

The output remains high and the relay remains on during a specific time interval adjusted by P1. This time depends also on C1 and can be changed according to the needs of the project.

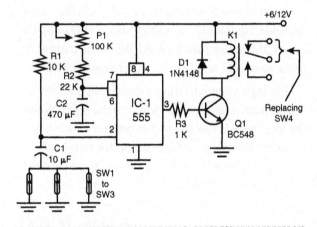

Figure 3.13.12 *A simple timer can be added to stop the elevator at each floor at a particular time interval.*

Parts List—Adding a Timer

- IC1 555 IC timer
- Q1 BC548 general-purpose *negative-positive-negative* (NPN) transistor
- D1 1N4148 general-purpose diode
- P1 100 kΩ trimmer potentiometer

R1 10 kΩ × 1/8-watt resistor (brown, black, orange)

R2 22 kΩ × 1/8-watt resistor (red, red, orange)

R3 1 kΩ × 1/8-watt resistor (brown, black, red)

C1 10 μF × 12-volt electrolytic capacitor

C2 470 μF × 12-volt electrolytic capacitor

K1 6-volt × 50 mA DPDT relay

PCB or solderless board, wires, solder, etc.

Reversing with an H-Bridge

Another circuit that can be used to change the direction of the motor at the end of course is the H-bridge. In an H-bridge, four transistors conduct the current in an alternating mode. In this way, depending on the transistors that are on, the current flows across the motor in opposite directions. Therefore, in the circuit shown in Figure 3.13.13, when transistors Q1 and Q4 are on, the motor runs forward. When the current flows across Q2 and Q3, the motor runs backward.

The logic system to trigger this circuit is a sensor (made from a reed switch) wired to the input. When the logic level is low, the motor runs forward, and when the input level is high, the motor runs backward. The transistor must be attached to a heatsink, and the power supply comes from a 6-volt source.

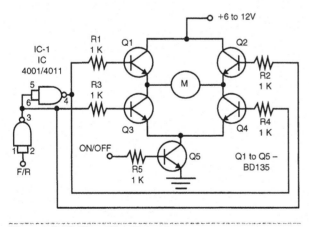

Figure 3.13.13 *The H-bridge.*

Parts List—Reversing with an H-Bridge

Q1 to Q5 BD135 NPN medium-power transistors

R1 to R5 1 kΩ × 1/8-watt resistor (brown, black, red)

IC-1 4011 or 4001 *complementary metal oxide semiconductor* (CMOS) IC

M1 6-volt DC motor (the one used in the project)

C1 1,000 μF × 12-volt electrolytic capacitor

Heatsink for the transistors, PCB or solderless board, wires, solder, etc.

Using a Flip-Flop

An automatic circuit for reversing the motor can be implemented using the 4013 CMOS D-type flip-flop. The circuit shown in Figure 3.13.14 closes and opens the relay when a command by the pulses is applied by the sensor connected to the input.

Starting from the moment the relay is off, a pulse is applied to the sensor, which triggers the 555 and produces a square pulse with a duration determined by R2 and C2. With this circuit it is important to debounce the sensors, thus avoiding erratic operation of the circuit.

The square pulse produced by the 555 is applied to the input of one of the D-type flip-flops of the 4013. This flip-flop toggles, driving the relay to the on state. To turn off the relay again, a new pulse must be applied to the input of the 555.

The RC network formed by C2 and R3 is necessary to start the circuit from the state where the relay is off when power is on.

Parts List—Using a Flip-Flop

IC-1 555 IC timer

IC-2 4013 CMOS IC, D-type flip-flop

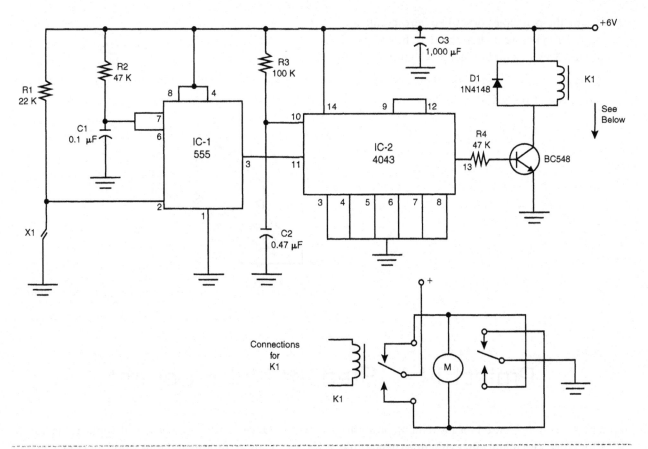

Figure 3.13.14 *Flip-flop to reverse the motor or to implement other functions on the elevator.*

Q1   BC548 general-purpose NPN silicon transistor

D1   1N4148 general-purpose silicon diode

R1   22 kΩ × 1/8-watt resistor (red, red, orange)

R2   47 kΩ × 1/8-watt resistor (yellow, violet, orange)

R3   100 kΩ × 1/8-watt resistor (brown, black, yellow)

R4   4.7 kΩ × 1/8-watt resistor (yellow, violet, red)

C1   0.1 μF ceramic or polyester capacitor

C2   0.47 μF ceramic or polyester capacitor

C3   1,000 μF × 12-volt electrolytic capacitor

K1   6-volt × 50 mA relay with contacts according to the application

X1   NO sensor

PCB or solderless board, wires, power supply, etc.

The Technology Today

Automatic elevators are found worldwide in nearly every building with two or more floors. They are generally controlled by microcontrollers performing intelligent functions. The same types of controls are also used in other automated routines such as production lines in factories and in domestic appliances with movable parts.

Ideas to Explore

- Starting from the basic project for an elevator, design a conveyor to transport small pieces, simulating a production line of a factory.

Section Three   The Projects   135

- Using the principles described here, create a transfer machine such as the one shown in Figure 3.13.15 simulating a production line.
- Design an elevator controlled by a microprocessor (PIC, 80C51, or other).

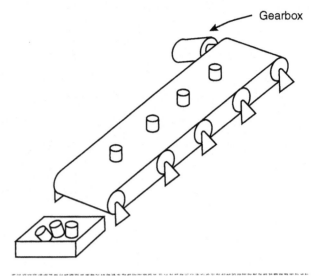

Figure 3.13.15 *A transfer machine using a rotating nut.*

Project 14 — Stepper Motor Control

Stepper motors may be used for movement, positioning, locomotion, and many other situations where the precise control of a shaft, lever, or any moving part is required. The purpose of this project is to provide the reader with a basic control for a four-phase stepper motor and with ideas for some additional circuits and applications in mechatronic projects.

The evil genius can use the stepper motor control to create robots, vehicles, and other automated devices. It also can be used to perform experiments in physics, as subjects with crossover themes, or simply to teach the reader how stepper motors function.

Objectives

- Provide the reader with the elements to build a simple stepper motor control.
- Explain how to use the stepper motor in robots and automatisms.
- Teach how a stepper motor works.
- Perform experiments in physics using a stepper motor.
- Develop ideas for practical applications of stepper motors in home appliances or mechatronic projects.

How a Stepper Motor Works

The operating principle of a stepper motor is not much different from that of a DC motor, such as the ones used in many other projects described in this book. Stepper motors are formed by coils and magnets and have a moving shaft that rotates when power is applied to the system. The difference is in the way the shaft is moved. In a stepper motor, the movement of the shaft occurs by applying power to different coils in a predetermined sequence (stepped).

The main applications for stepper motors are not ones that require spinning but ones with fine control requirements where the number of steps per second is important. Another important feature of stepper motors is that they can hold their position and resist turning.

Stepper motors are found in many sizes and shapes according to the application and the power to be delivered. Figures 3.14.1 and 3.14.2 show common four-phase stepper motors.

Figure 3.14.1   *A four-phase stepper motor.*

Figure 3.14.2   *Another four-phase stepper motor.*

Stepper motors can be found in three basic types: permanent magnet, variable reluctance, and hybrid. The way the windings are organized inside the motor determines how it works. The most common type is the four-phase stepper motor, ghich has the four coils organized as shown in Figure 3.14.3.

Because this kind of motor has pairs of windings with a common connection, it can be easily identified by the six wires. In normal operation, the common wires are connected to the positive rail of the power supply. The other wires are connected to the ground for short time periods in a sequence that depends on the movement you want to produce.

Each time a winding is energized, the motor advances by a fraction of a revolution. To make the shaft turn forward or backward in complete revolutions, you must apply pulses to the windings in a pre-determined sequence. Figure 3.14.4 shows the normal sequence of commands to make a stepper motor rotate.

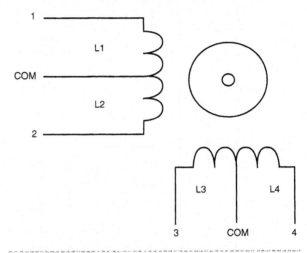

Figure 3.14.3   *The symbol and coil organization of a four-phase stepper motor.*

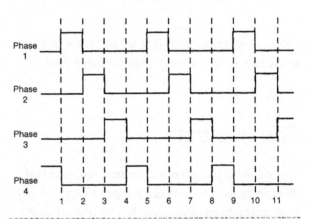

Figure 3.14.4   *Sequence of pulses applied to a stepper motor.*

Another way to control the movement of a stepper motor is by on/off pulses in a sequence such as the one shown in Figure 3.14.5.

A common stepper motor can be powered from 5-volt, 6-volt, or 12-volt sources. The current drain depends on the type or the power to be delivered to the shaft (torque).

The motors that can be found in many appliances and bought in components stores normally are rated to currents in the range of 100 mA to 2 A. For our project, we recommend the use of motors up to 500 mA in the basic version and up to 2 amps as you change some components.

The pulse rate of a stepper motor is limited. Common types have the speed limited to 200 pulses per second, or approximately two or three turns per second. This means that stepper motors are low speed

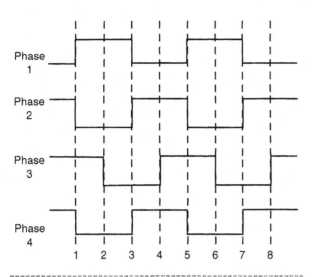

Figure 3.14.5 *Controlling a motor by an on/off sequence of pulses.*

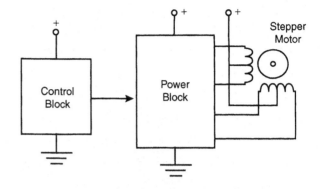

Figure 3.14.6 *Block diagram for the project.*

and are low-torque units that are better for tasks where precision is more important than power.

The Project

The following paragraphs describe a simple circuit that can be used to control a stepper motor and give some suggestions of projects the evil genius can create using one or two of those motors.

A stepper motor must be controlled by pulses applied in sequence to the windings. Therefore, it is necessary to create a circuit that generates the pulses for external commands such as switches, joysticks, sensors, or even the parallel port of a computer.

So what we are going to describe here is a project formed by two blocks: (1) a power block that can drive stepper motors up to 500 mA and (2) control blocks that generate sequences for the movement of the stepper motor. Figure 3.14.6 shows the block diagram for the project.

Notice that we will give three options for control blocks, which can be used according to the application:

- **Manual operation—using switches or a joystick**: The pulses can be generated by switches or a joystick, allowing the operator to place the shaft of a stepper motor exactly in the position he or she wants. A remote-controlled robotic arm or other automation is a project that can be designed using this arrangement.

- **Sequential operation—using a pulse generator**: In this case, a circuit produces pulses that make the shaft rotate. The circuit can be used in automated devices where the pulses come from sensors.

- **Automatic operation—using an oscillator**: A free-running oscillator is used to rotate the motor. The frequency of the oscillator determines the speed of the shaft and the sequence of the direction. In this application, the motor can be used to produce movement. Robots and automated devices can be designed.

How It Works

The power block is formed by four *negative-positive-negative* (NPN) medium-power BD135 transistors coupled directly to the stepper motor windings. Each transistor acts as a switch, turning on when a positive voltage is applied to its base.

The transistors have high gain, needing only 0.5 mA to drive the motor winding that has currents up to 500 mA. This means that the output from a sensor, a logic circuit, and even a simple computer interface can be used to drive this stage. Because the loads controlled by the transistors are independent, motors rated to voltages between 5 and 12 volts can be used. Of course, the transistors must be mounted on heatsinks.

The switched block for manual operation is formed by on/off switches, sensors, or a joystick in an arrangement that can be used to produce the sequences needed for the application. Because the power block is sensitive, this block doesn't need an amplification stage.

The second control block is a sequencer that produces the necessary sequence to rotate the motor shaft forward or backward according to the pulses produced by the sensors or switches. It is an ideal configuration to show how a stepper motor works. You can also use it in several types of automatic devices.

The third block includes an audio oscillator used to generate steps in sequence and make the motor rotate forward and backward. The oscillator has the frequency adjusted externally by a potentiometer, but other methods of control can be used. An interesting idea is to use resistive sensors. The speed at which the motor rotates can be changed by the amount of light falling onto a *light-dependent resistor* (LDR) or by the temperature changes registered on a sensor like a *negative temperature coefficient* (NTC) resistor.

The circuit can be mounted on a small *printed circuit board* (PCB), and the transistor must be attached to heatsinks. Figure 3.14.8 suggests a pattern for a printed circuit where this circuit can be mounted.

Equivalent transistors for the BD135 that can be used as replacements are as follows:

- **BD137, BD139**: Up to 500 mA and no modifications are needed in the project.
- **TIP31, A, B, or C**: Up to 2 amps and the transistor must be inverted, as the terminal placement is different from the BDs. Figure 3.14.9 shows this placement.
- **TIP110, TIP111, TTP112**: Darlington transistor controlling up to 1 amp. The base resistor can be increased up to 10 kΩ; this version is more sensitive. Terminal placement is the same as for the TIP31.

The power supply can provide the voltage necessary to drive the stepper motor. Batteries and power supplies from the AC power line can be used.

How to Build

Figure 3.14.7 shows the basic configuration of the power stage using four NPN transistors.

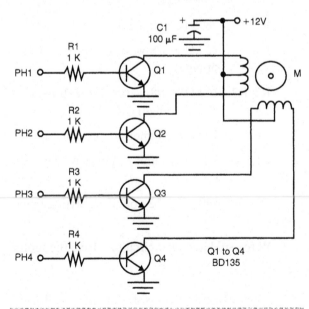

Figure 3.14.7 *Power stage using NPN transistors.*

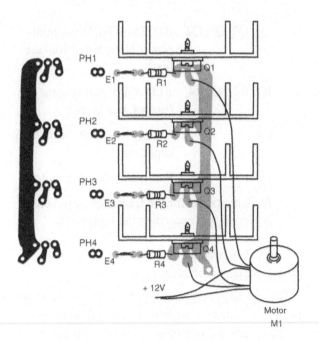

Figure 3.14.8 *PCB for the power stage.*

Section Three   The Projects   139

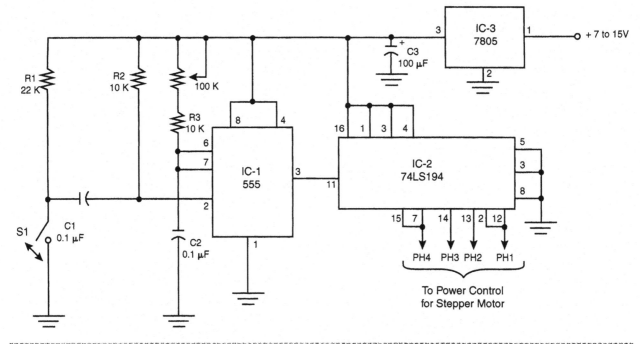

Figure 3.14.9 *Manual control for the stepper motor using on/off switches, a joystick, or sensor.*

Parts List—Power Circuit

Q1, Q2, Q3, Q4	BD135 or TIP31 medium-power NPN transistor (see text)
R1, R2, R3, R4	1 kΩ × 1/8-watt resistor (brown, black, red)
C1	100 μF × 12-volt electrolytic capacitor
M	Four-phase stepper motor

PCB, heatsinks for the transistor, wire, solder, power supply according the motor, etc.

Figure 3.14.9 shows the circuit for the control using switches, on/off sensors, or a joystick.

This circuit generates pulses in sequence each time the sensor is closed. Pulses of a duration determined by the adjustment of P1 are produced, making the motor advance one step each time the circuit is activated.

Because the current needs of the circuit are very low, reed switches can be used as sensors. Adjust P1 to have the necessary time to drive the motor.

Parts List—Stepper Motor Control

S1	On/off (SPST) switch or sensor
IC-1	555 *integrated circuit* (IC) timer
IC-2	74LS194 IC transistob-transistor logic (TTL)
IC-3	7805 IC voltage regulator
R1	22 kΩ × 1/8-watt resistor (red, red, orange)
R2, R3	10 kΩ × 1/8-watt resistor (brown, black, orange)
P1	100 kΩ trimmer potentiometer
C1, C2	0.1 μF ceramic or polyester capacitor
C3	100 μF × 12-volt electrolytic capacitor

Wires, solder, power circuit to drive the motor, etc.

The third control circuit is shown in Figure 3.14.10. This circuit generates a sequence of pulses that makes the stepper motor rotate. The sequence can be inverted to make the motor run backward, and the

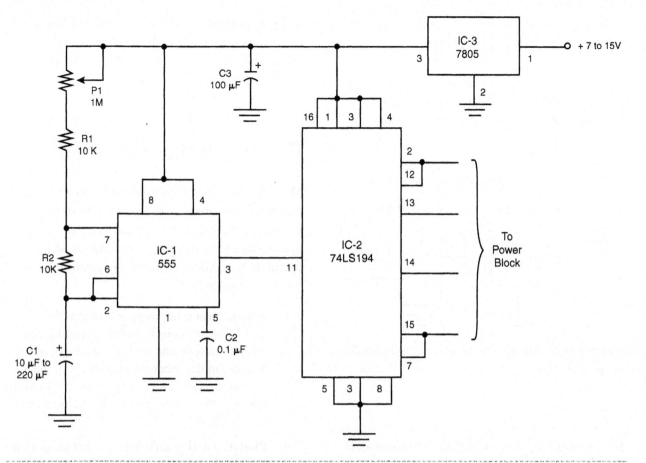

Figure 3.14.10 *Sequence generator for the stepper motor control.*

speed can be changed by altering the frequency of the oscillator.

The capacitor determines the speed range of the motor. In demonstrations or low-speed applications, you may use higher values for C1. Values between 10 μF and 220 μF are recommended, but if you want higher speeds you may use lower capacitances for C1. Values between 0.047 μF and 0.47 μF are recommended.

The switches between the logic circuit and the oscillator allow the circuit to be stalled by an external control. This switch is important when using the circuit in demonstrations. The inverse sequence used to make the motor rotate backward is generated, inverting the counter sequence.

The logic block can be powered by 5- to 15-volt power supplies from the same source that runs the motor or from another source. If different sources are used, they must have a common ground point.

Parts List—Sequencer

IC-1	555 IC timer
IC-2	74LS194 IC TTL
IC-3	7805 IC voltage regulator
R1, R2	10 kΩ × 1/8-watt resistor (brown, black, orange)
P1	1 MΩ potentiometer
C1	10 to 220 μF × 12-volt electrolytic capacitor
C2	0.1 μF ceramic or polyester capacitor
C3	100 μF × 12-volt electrolytic capacitor

Power module, PCB or solderless board, wires, power supply, etc.

Section Three The Projects

Testing and Using

Figure 3.14.11 shows how to plug the sequencer block into the power block in order to have a control for the stepper motor and make the main operational tests.

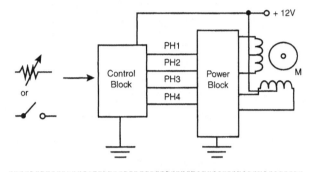

Figure 3.14.11 *Testing the stepper motor control in the basic version.*

Take care to connect the stepper motor, observing the position of the wires according to their colors. If they are inverted, the motor will not rotate. Power on the circuit and close S1 to generate the sequence of pulses. By adjusting P1, the motor will rotate according to the direction programmed by S2.

If your motor doesn't rotate but simply changes the position of the shaft randomly, it is because (1) the wires are not correctly connected to the circuit or (2) your motor has a different sequence for operation. If you want to use the motor with other configurations (switches, sensors, a joystick, etc.), you must test that circuit now.

The next step is to find an application for your control. Figure 3.14.12 shows how you can use the circuit to control the movement of a mechatronic head using the sequencer and the power circuit. You can add effects to the head such as a sound circuit to make it speak and move its mouth using *shape memory alloy* (SMA) or solenoids.

Exploring the Project

Many mechatronic projects can be based on a stepper motor. The precise movement of the stepper motor's shaft can be used in position-critical parts of mechatronic projects such as arms, robots, and automated devices. Additional ideas are given in the following paragraphs:

- **Robotic arm using stepper motors:** Figure 3.14.13 shows a mechatronic arm using two stepper motors, one for the movement in the X axis and the other for movement in the Y axis. Depending on the power needed to move the arm, the stepper motor should be coupled to a gearbox.

- **Plotter:** Another idea for a project using stepper motors is a plotter, shown in a simplified configuration in Figure 3.14.14. Two motors control the movement of a pen in the X and Y axes. The stepper can be controlled by a computer, and a program can be prepared to transfer figures to the plotter. It is a great project requiring the skills of the designer because precision is very important to achieve good results.

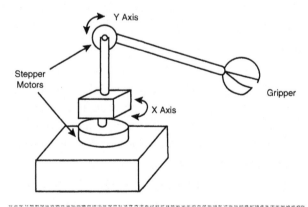

Figure 3.14.13 *Robotic or mechatronic arm using two stepper motors.*

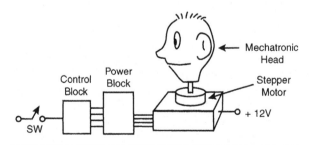

Figure 3.14.12 *Controlling the movement of a mechatronic head.*

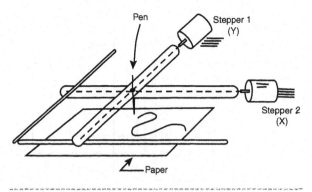

Figure 3.14.14 *Basic idea for a plotter.*

- **Controlling a camera:** One or two stepper motors can be used to control the movement of a video camera, as shown in Figure 3.14.15. If one stepper motor is used, the camera can sweep across a place in a horizontal movement. If two motors are used, you can move the camera up and down and left and right.

- **Controlling a laser beam:** A laser pointer or a laser module can be controlled by one or two stepper motors generating figures in a screen as described in Project 10.

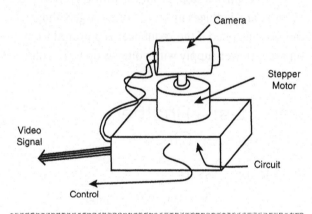

Figure 3.14.15 *Controlling a video camera.*

Cross Themes

Stepper motors are precise tools for experiments in physics. A teacher can create many devices to show in the classroom, performing experiments in classes that provide crossover subject matter, such as optics, mechanics, and other sciences. A few crossover ideas are given here:

- **Newton's color wheel:** The same experiment described in the touch-controlled motor section (Project 12) can be performed using a stepper motor. To use a stepper motor in this experiment, the reader must adjust only the speed to have the best luck mixing colors that result in white.

- **Standing waves:** The apparatus shown in Figure 3.14.16 can be used to produce standing waves in a wire. Adjust P1 to have the speed that generates the desired number of complete waves in the wire.

- **Gears and movement:** The fine control of speed and the shaft position of a stepper motor can be used to perform experiments with gears and movement in physics classes.

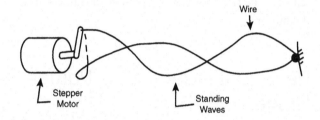

Figure 3.14.16 *Producing standing waves.*

Additional Circuits and Ideas

Many configurations can be used to control stepper motors. Some of them are suggested in the following sections.

Using a Darlington Transistor

Darlington transistors such as the TIP110/TIP111/TIP112 can be used in the power stage, as shown by the circuit in Figure 3.14.17. These transistors can control currents up to 1.25 amps, and others in the same series of TIP transistors can control higher currents (up to 5 amps). These must be installed on heatsinks and powered for a large power supply.

Section Three   The Projects   143

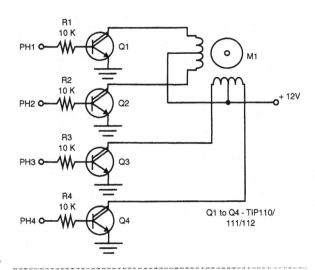

Figure 3.14.17 *Using Darlington transistors.*

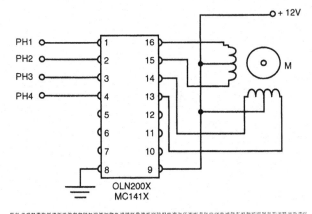

Figure 3.14.18 *Using the ULN2001A or MC1411 IC.*

The great advantage when using Darlington transistors is the need for lower-input currents to control the motor.

Parts List—Darlington Power Stage

- Q1 to Q4 TIP110/111/112 silicon NPN Darlington power transistors
- R1 to R4 10 kΩ × 1/8-watt resistor (brown, black, orange)

Heatsinks for the transistors, wires, PCB or solderless board, solder, etc.

Using an Integrated Circuit (IC)

The ULN2001A or MC1411 are ICs specially designed to control high-power loads of solenoids, motors, relays, and stepper motors. These ICs match with *complementary metal oxide semiconductor* (CMOS) logic and TTL, making it easier to design control circuits.

Figure 3.14.18 shows an IC in a typical application controlling a stepper motor. This circuit can replace the transistor stage described in our basic version of the control and can drive motors draining currents up to 500 mA.

Another Sequencer Circuit

Figure 3.14.19 shows another sequencer circuit that can be used in projects with stepper motors. This circuit is a simple 1-to-4 counter using the 4017. The output stage can use the ULN2001A or MC1411, and the pulse and sequence generators are the same described in our basic project.

It can control a stepper motor with currents up to 500 mA and voltages up to 12 volts. Large voltages can be controlled if the sequencer is powered from a separate power supply with voltages up to 12 volts.

Parts List—Another Sequencer Circuit

- IC-1 555 IC
- IC-2 4017 CMOS IC counter
- IC-3 ULN2001A or MC1411 power driver
- X1 6- or 12-volt stepper motor
- P1 1 MΩ potentiometer (linear or logarithmic)
- R1, R2 10 kΩ × 1/8-watt resistor (brown, black, orange)
- C1 47 μF × 12-volt electrolytic capacitor
- C2 470 μF × 16-volt electrolytic capacitor

Power supply, solderless board or PCB, wires, etc.

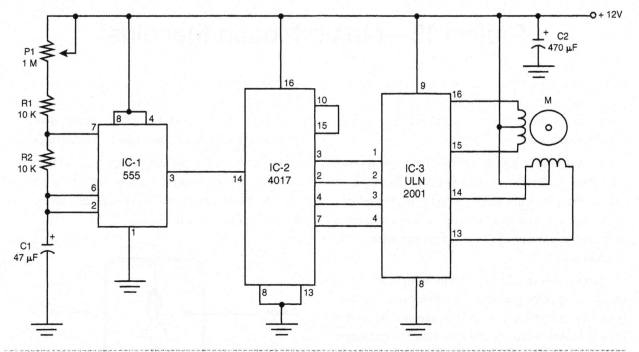

Figure 3.14.19 *A sequencer circuit using the 4017 CMOS IC.*

The Technology Today

You will find stepper motors in many domestic appliances. Computers, for instance, use stepper motors in many of their peripherals. The constant speed of a disk drive or a hard disk is maintained by stepper motors. CD and DVD players also use stepper motors to rotate the disks because the speed must be controlled with high accuracy.

The movement of the head of your printer is also controlled by a stepper motor. The stepper motor, controlled by software, can place the head exactly where it must produce a dot.

Ideas to Explore

- Design automated devices using a stepper motor.
- Find information on the Internet about two-phase stepper motors.
- Design a mechatronic arm that can be precise enough to handle small objects using stepper motors.

Project 15—Magic Motion Machine

Mobile figures found in many decorative objects often are operated by energy from cells. Many watches have figures that dance and even play music to indicate the hour.

Imagine a figure that moves randomly and is powered from an electronic source but controlled by a circuit. The aim of this project is to build a figure and circuit that can be used in decoration, animatronics (a branch of mechatronics), or as a demonstration in mechatronics.

The circuit is simple, uses common parts, and has an operating principle used in another project of this book (for example, see the galvanometer described in Project 5). An electronic circuit is added to produce the motion of the toy or other figure that the reader can create.

Objectives

- Mount a simple circuit to control a mobile figure.
- Make a circuit that produces random movements.
- Design automatic and mechatronic projects using the same operational principle.

The Project

This project should be seen as an upgrade of Project 5, the experimental galvanometer. We will add an electronic circuit to produce electric pulses and replace the needle from the galvanometer with a mobile figure such as a ballet dancer, a clown, or another figure.

The project is powered from AA cells, which should have a good life when powering this circuit. The project also includes a *light-emitting diode* (LED) that blinks, giving the rhythm to the mobile figure.

You can place this circuit in a box, as suggested by Figure 3.15.1, and have a homemade, magic, decorative object—an important point to consider if you intend to demonstrate your skills in mechatronics.

The name of this project, "magic," comes from the fact that no power source to move the object is visible. The movement is created by magnetic induction, a type of power we can't see.

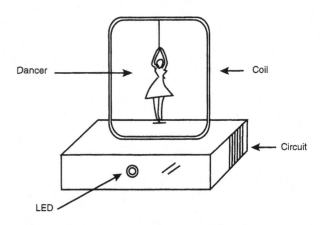

Figure 3.15.1 *Decorative object made with the magic motion project.*

How It Works

The main block of our project is a 555 *integrated circuit* (IC) configured as an astable multivibrator. This circuit produces intervaled pulses that drive the mechanical part of the project.

In this configuration, the duration and interval between pulses depend on four components of the circuit, as shown in Figure 3.15.2. The time interval in which the output remains in the high state depends on four components of the circuit: the resistance Ra (R1 + P1), the resistance Rb (R2), the capacitance C (C1), and the following formula, which ties them all together:

$$th = 0.693 \times C \times (Ra + Rb)$$
$$tl = 0.693 \times Rb \times C$$

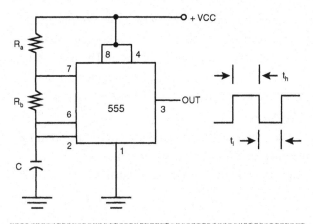

Figure 3.15.2   *The astable 555.*

where:

th = the time interval of the output high in seconds (s)

tl = the time interval of the output low in seconds (s)

C = the capacitance in farads (F)

Ra, Rb = the resistances in ohms (Ω)

Note that, because Ra + Rb is always higher than Rb, the time high is always greater than the time low. Note also that Ra is given by R1 plus the adjustment of P1, so it is variable.

As we want the inverse in the load, the solution is to excite it with the low pulses of the 555. This can be accomplished by using a *positive-negative-positive* (PNP) transistor to drive the load, and the transistor conducts with the low pulses of the 555. Therefore, as Figure 3.15.3 shows, the circuit applies pulses to the load in regular intervals, determined by the adjustment of P1.

The transistor is a medium-power type that drives a load creating an intermittent magnetic field. This field acts on magnetic objects such as metal blades, needles, or even other magnets (turning them). Because the field is not strong, the objects must be free to move, which requires a high level of skill in the builder as they will need to be suspended in some way.

For the best performance, the critical point is to find a pulse rate that matches the apparent natural movement of the object. The pulse rate shouldn't be too fast for a ballet dancer or too slow for a runner. Tests must be done to find the correct frequency and even alter the values of R2 in the range of 15 kΩ to 47 kΩ.

How to Build

The mechatronic project can be divided in two parts for the purposes of construction: (1) the electronic circuit and (2) the mechanic part.

Electronic Circuit

Figure 3.15.4 shows the electronic circuit for the pulse generator using the 555 IC. Figure 3.15.5 shows how the circuit can be mounted using a solderless board. Of course, the reader is free to use other mounting technology.

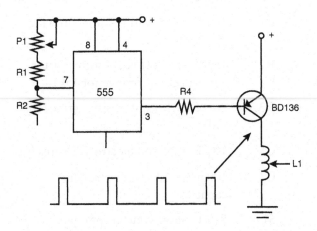

Figure 3.15.3   *Waveform of the signals applied to the load.*

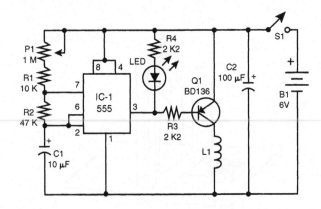

Figure 3.15.4   *The schematics for the pulse generator.*

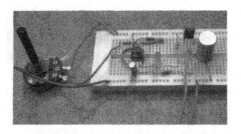

Figure 3.15.5 *The circuit can be mounted on a solderless board for experimenting.*

The builder must take care with the position of the polarized components such as the IC, electrolytic capacitors, transistor, and LED. If the TIP32 is used as an equivalent replacement for the BD136, the reader must remember that the wire leads have different identifications. The circuit is powered from AA cells, but the evil genius can also use a power supply from the AC power line. Types between 3 and 6 volts can be used.

Mechanical Part

The mechanical part is made up of the coil and the mobile figure. In our basic version, the coil is formed by 50 to 100 turns of any enameled wire with gauges between 28 and 32 AWG. The form is a cardboard box, as shown in Figure 3.15.6.

It is important to remove the enamel at the ends of the coil to allow electric contact when attaching to the screw on the terminal strip or soldering to the circuit.

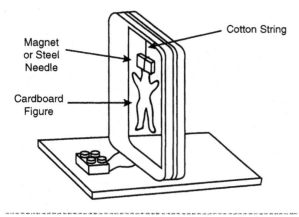

Figure 3.15.6 *The mechanical part of the project.*

The mobile object is hung by a thin cotton string or wire. The mobile object will have a small magnet or a small metal object attached to it. The metal must be a type that can be attracted by magnets such as iron or steel. Magnets do not attract aluminum or copper.

The builder must also find the correct position for the magnet. It must be positioned in a way that the magnetic field of the coil acts to turn it.

Test and Use

Power the circuit on and observe the movement of the mobile figure and the blinking of the LEDs. Each time the LED lights up, the object should make some movement, as both are powered by the same force. If not, take a look in the contacts of the coil. If no problem is found, adjust P1 to create the desired movement of the mobile figure.

Parts List—Basic Magic Motion Machine

IC-1	555 IC timer
Q1	BD136 silicon medium-power PNP transistor
LED1	Common LED (any color)
P1	1 MΩ trimmer potentiometer or common linear or logarithmic potentiometer
R1	10 kΩ × 1/8-watt resistor (brown, black, orange)
R2	15 kΩ to 47 kΩ × 1/8-watt resistor (brown, green, orange for 15k)
R3, R4	2.2 kΩ × 1/8-watt resistor (red, red, red)
C1	10 μF × 12-volt electrolytic capacitor
C2	100 μF × 12-volt electrolytic capacitor
S1	SPST on/off switch (optional)
B1	6 volts of power (4 AA cells and holder)

L1   Coil (see text)

Printed circuit board (PCB) or solderless board, box, magnet, solder, wires, etc.

Cross Themes

This project can be used in elementary and high school science and physics courses to show how energy transforms and manifests itself. Some ideas to link the theme with the project are given here:

- Study of energy conservation
- Production of magnetic fields by electric currents
- Study of the perpetual movement (explains why energy can't be created or destroyed but only transformed)

Exploring the Project

Some changes in the basic version can help the reader create interesting decorative objects or projects for demonstration. The way the magnetic field is applied to the moving figure can be changed. Figure 3.15.7 shows how a mobile athlete can do exercises with loops, using the basic circuit.

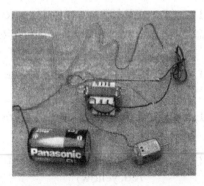

Figure 3.15.7 *Another version of the magic motion machine.*

The coils must be wound in the correct direction because, when the currents flow, the magnetic field of each coil repels the other, creating the force that makes the movement of the athlete. Tests must be made to find the correct number of turns of each coil and the size of the cardboard athlete.

Additional Circuits and Ideas

Generating pulses for the magic motion project can be done in many ways. Some additional circuits are shown in the following section.

Another Pulse Generator

The circuit shown in Figure 3.15.8 uses a 4093 IC as a very low freauency generator that drives the BD136, as in the original project.

The diodes determine the current's path for the charge and discharge of the capacitor. Therefore, R2 determines the time at which the output of the circuit is low, and the adjustment of P1 and R1 determines the time of the output high. Because the three other gates of the 4093 are wired as inverters, the result is that the coil is energized with the pulses high, and the interval between them is determined by the pulses low.

Parts List—A Pulse Generator Using a 4093 IC

IC-1	4093 *complementary metal oxide semiconductor* (CMOS) IC
D1, D2	1N4148 general-purpose silicon diode
Q1	BD136 silicon medium-power PNP transistor
P1	1 MΩ trimmer potentiometer or common potentiometer (linear or logarithmic)
R1	10 kΩ × 1/8-watt resistor (brown, black, orange)

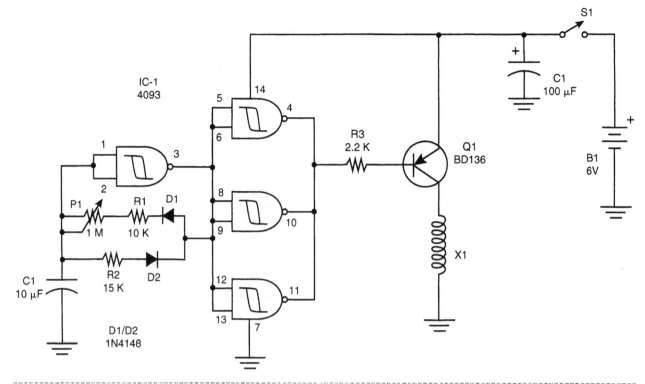

Figure 3.15.8 *Circuit using the 4093 IC.*

R2	15 kΩ × 1/8-watt resistor (brown, green, orange)
R3	2.2 kΩ × 1/8-watt resistor (red, red, red)
C1	10 μF × 12-volt electrolytic capacitor
C1	100 μF × 12-volt electrolytic capacitor
S1	SPST switch (on/off)
B1	6 volts of power (4 AA cells and holder)
X1	Coil (the same as in the basic project)

PCB or solderless board, plastic box, wires, solder, etc.

A Random Pulse Generator

Figure 3.15.9 shows an interesting circuit that produces random pulses made by three different oscillators. R2, R4, and R6 determine when each of the oscillators is high. Therefore, combining the different times, the pulses are produced in a random mode, as shown in Figure 3.15.10.

Parts List—Random Pulse Generator

IC-1	4093 CMOS IC
Q1	BD139 or equivalent medium-power *negative-positive-negative* (NPN) silicon transistor
D1 to D9	1N4148 general-purpose silicon diodes
R1, R3, R5	15 kΩ × 1/8-watt resistor (brown, black, orange)
R2	220 kΩ × 1/8-watt resistor (red, red, yellow)
R4	330 kΩ × 1/8-watt resistor (orange, orange, yellow)
R6	470 kΩ × 1/8-watt resistor (yellow, violet, yellow)
R7	2.2 kΩ × 1/8-watt resistor (red, red, red)

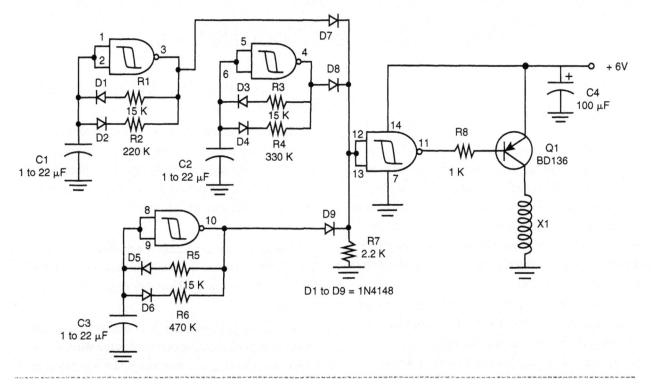

Figure 3.15.9 *This oscillator produces random pulses.*

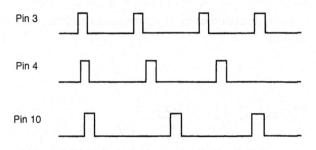

Figure 3.15.10 *A random oscillator using the 4093 IC.*

R8	1 kΩ × 1/8-watt resistor (brown, black, red)
C1, C2, C3	1 μF to 22 μF × 12-volt electrolytic capacitors (according to the frequency)
C4	100 μF × 12-volt electrolytic capacitor
X1	Coil (as in the basic project)

PCB or solderless board, power supply, wires, solder, etc.

The Technology Today

The best example of an application for the principle explored in this project is the magic motion clocks made mainly by Japanese industries and popular in every part of the world. These clocks come in a variety of sizes and degrees of sophistication, presenting one or more mobile figures. In many homes, they are used as the primary decorative object.

Ideas to Explore

Other projects can be designed, starting from the same principles described here:

- Ampere's Force law is studied in physics. This is the law used in this experiment. You can find more on the Internet.

- Design an electric motor based on the principles described here. You'll find many ideas on the Internet.

Project 16—Test Your Nerves

Introduction

Two versions of this project will be given. The simplest version uses only a few connections and needs no chassis or other special resources. It is the ideal project in which to learn how to solder, making it perfect for students aged 11 to 16. They can test their manual skills at the same time they are being introduced to the operating principle behind transformers and other components.

Of course, the evil genius who is learning how to solder can use this project as the first step to more complex inventions, can upgrade this project to build the second version, and even play some jokes on friends or family. Giving an electric shock to your friends can be a very good joke (if done safely), and you can use this project and your skills. (Please read the caution in the text.)

The basic idea is to increase the voltage of one or two cells to a value great enough to cause a shock. Because the current is limited to few milliamperes and the shock doesn't last more than few seconds, we consider the discharge inoffensive and safe.

Objectives

- Generate a high enough voltage from cells to apply a shock to friends.
- Learn how a transformer works.
- Learn how to solder.
- Make some experiments with high voltage.

The Project

This circuit will put your manual skill to a good test. As shown in the schematic diagram of the basic version, the circuit has a small loop that you must navigate around a wire. The object of the game is to guide the loop over the winding and weaving course without touching the wire.

Any misjudgment or quiver of the hand and the ring will contact the weaving wire, turning on the circuit and producing a slight shock! The skill required to play the game depends largely on the size of the loop and the degree of twisting and turning in the wire.

If used as a game, scoring is a matter of counting the number of times the player is shocked! The person with the lowest total number of shocks (or no shocks) wins.

The circuit is powered by a D cell, as the current drain is high when the loop touches the wire. When not in use, the ring and wire should be separated or the cell will run down in a short period of time.

The basic version is very simple and can be mounted in 1 or 2 hours, but we will give a second improved version using more resources and mechatronics.

How It Works

The core of the project is a small transformer having a primary coil rated to the AC power line (or 117 VAC) and a secondary coil rated to 6 to 9 volts, with currents ranging from 250 to 500 mA.

In the normal operation, the transformer receives 117 VAC when its primary coil converts it to a low voltage (6 to 9 volts according to the unit). But in this project we show that the transformer can be used in a different way: inverting. Applying a low voltage (6 to 9 volts) from a cell to the secondary coil, the voltage rises, producing high-voltage pulses in the primary coil.

Because the voltage applied is not in the form of a sinusoidal wave, as found in the power supply line, but consists of sharp pulses, the voltage in the primary coil can rise to values between 100 and 300 volts. Of course, the pulses can cause a severe shock, but because the duration of the pulses is very short

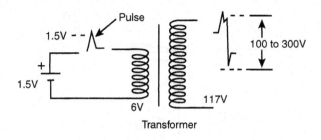

Figure 3.16.1 *A transformer is used to raise the voltage of a cell.*

and the current very low, they are not harmful. Figure 3.16.1 shows what happens.

Notice that the induction process that converts the low voltage to a high one in the transformer is dynamic. This is the mechatronic part of the basic version of this project. It occurs only at the moment when the wire touches the loop and when the wire disconnects from the loop. This also explains why we can't have high voltage all the time in the secondary coil even if we keep the wires together. A change in the current is required to produce the instantaneous high voltage, and this is reached only in the moments when the wires are touched and separated.

In the second version, we are simply going to use a small DC motor to act as a variable high-speed switch. Small DC motors have brushes that act as switches when they rotate. They open and close the coils producing a variable current in the circuit. The idea is to place a small motor in series with the transformer, helping in the generation of a variable current.

Premounting Experiments

Before mounting, you can show the operating principles of a "test your nerves" project with some simple and interesting experiments.

Shocking!

You can show how a transformer can be used to raise the voltage of a cell and cause electric shocks. You can also show that you need a closed circuit to have an electric current.

Parts List—Shocking

T1 Any transformer with a 117 VAC coil and a low-voltage secondary coil (4.5 to 9 volts)

B1 1.5 volts of power (a D cell)

The Experiment—Invite your friends or relatives to hold hands with one another forming a circle, as shown in Figure 3.16.2. Choose certain people on opposite sides of the circle to hold the wires of the high-voltage coil of the transformer in their hands.

By touching the wires of the primary coil to the cell, a high-voltage pulse will cause a shock in *all* the people in the circle. This is a perfect opportunity to explain the following:

- The current is the same in all persons and therefore the shock is the same in each person.
- What a closed circuit is.

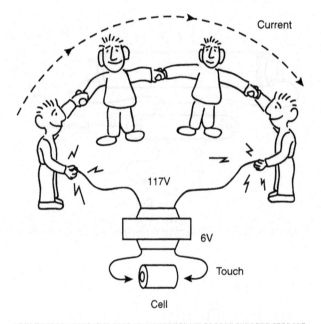

Figure 3.16.2 *The shocking circle.*

Lighting a Fluorescent Lamp

You can show the induction of high voltage by the transformer lighting a fluorescent lamp.

Parts List—Lighting a Fluorescent Lamp

T1 Any transformer with a 117 VAC coil and a low-voltage secondary coil (the same as the ones used in the previous experiment and in previous projects)

B1 1.5 volts of power (D cell)

A small file and wires

The Experiment—Notice that the lamp flashes only when you touch the wires to the cell. The use of a file can help you produce fast changes in the current flow, thus producing more powerful flashes in the lamp. Any fluorescent lamp can be used in this experiment. Figure 3.16.3 shows the experiment.

This project presents a perfect mechanism for explaining the following:

- Why the lamp flashes only when the current changes
- Why the lamp will not glow with 100 percent of its power

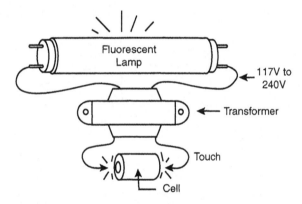

Figure 3.16.3   *Powering a fluorescent lamp from a cell.*

How to Build

This circuit is very simple. No special skills are needed to build it. The construction of the two versions is described in the next lines.

Test Your Nerves Version 1

The schematic for the device is shown in Figure 3.16.4. The heart of the project is the transformer that

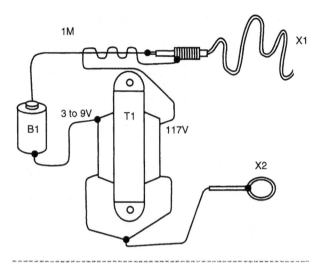

Figure 3.16.4   *The schematics for version 1.*

converts 1.5 DC volts into high-voltage pulses up to 300 volts in some cases! The mechanical view of the mounting is shown in Figure 3.16.5.

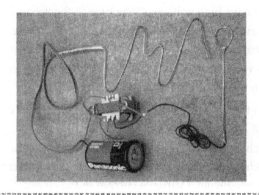

Figure 3.16.5   *All parts of the project.*

All the components can be housed in a small plastic box. The wires to the loop and the weaving wire can be 2 or 3 feet long to prevent pullouts as the player pulls back when reacting to the shock.

Observe the isolation between the points where the player touches and the weaving wire. The two wires that connect should be the twisted or parallel type. Use different colors to prevent inverting the connections.

The weaving wire is made from a bare copper wire with a small portion kept isolated or covered. In this isolated part, a length of thin, bare wire forms the handle. The loop is also made from a length of bare wire and doesn't need a handle because low-voltage and high-voltage lines are common in this part of the circuit.

T1 is any small transformer with a 117 VAC primary coil and a secondary coil ranging from 3 to 9 volts. Current drain can range from 250 to 500 mA when the wire touches the loop.

Testing and Using

Testing is very simple. Find a volunteer to try the test. By the volunteer's expression when he or she touches the loop to the weaving wire, you will know if your project works or not. But if you don't find a volunteer, you have no alternative. You must make the test yourself.

When using the device, do not let the loop contact the heavy wire. The current flowing doesn't induce high voltage (because it is necessary to change the intensity), and the cells will run down quickly.

Caution! Don't use this circuit on persons who have pacemakers or heart disease!

Parts List—Test Your Nerves Version 1

- T1 Any transformer with a 117 VAC coil and a low-voltage secondary coil (see text)
- B1 1.5 volts of power (D cell)
- X1 Loop (see text)
- X2 Weaving wire (see text)
- Wires, solder, plastic box, etc.

Test Your Nerves Version 2

The schematic diagram for the second version is shown in Figure 3.16.6. In this version, we add a small DC motor in series with the secondary coil of the transformer. When the motor is on, the brushes produce fast pulses that induce high voltage in the transformer. The diagram of the final project is shown in Figure 3.16.7.

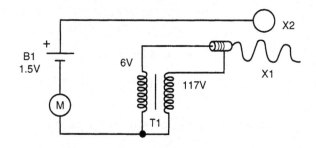

Figure 3.16.6 *Schematic for the second version.*

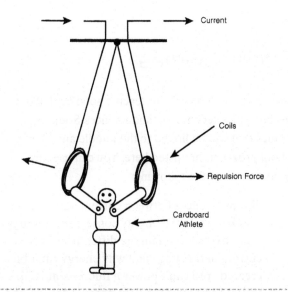

Figure 3.16.7 *The second version can be built on a small piece of wood or plastic.*

A heavy disk of any material is attached to the shaft of the motor to reduce its speed and then help in the production of the pulses to the transformer.

Testing and Using

The procedures for testing are the same as described in version 1. The main difference is that if you keep the loop touching the weaving wire, the current across the circuit is not so high because it is limited by the motor.

Parts List—Test Your Nerves Version 2

- T1 Any small transformer with a 117 VAC primary coil (5 to 9 volts), and a secondary coil with currents ranging from 200 to 500 mA

M   Any small DC motor

B1 3 volts of power (two C or D cells)

X1 and X2 Loop and wavy wires (as in version 1)

Plastic box, wires, solder, etc.

Cross Themes

This project can be used to teach a good deal about mechatronics, technology, and sciences. Some cross themes that can be linked to the middle and high school programs are listed here. You can use this project to study the following:

- The differences between current and voltage, also known as the energy conservation principle (physics). Explain the differences between current and voltage and why energy can't be created. You can't power a fluorescent lamp to 100 percent of the light because the energy sourced by a cell isn't enough.

- Electric induction and how a transformer works (physics). Explain how a transformer works, and how energy can be transferred from one place to another through a magnetic field.

- Electric shock (biology). Find information about the effects of the electric current on living beings, how current can be dangerous, and how it can be used to help persons with diseases (for example, in heart stimulation).

- Ionization of a neon or fluorescent lamp (chemistry). Explain how gases are ionized by high voltages and how the phenomenon is used in several appliances.

- Testing for stress (human sciences). Use the test-your-nerves device with persons submitted to several levels of stress. Show how the number of successes or failures can be a function of existing stress.

Additional Circuits and Ideas

The basic project can be improved with some variations or by using an electronic circuit to increase the voltage. The circuits that can be used with this task are called *inverters* and are found in many appliances.

Using a Reed Switch

Figure 3.16.8 shows how a magnet attached to the shaft of a motor can trigger a reed switch, thus opening and closing the circuit and generating high voltages across the transformer.

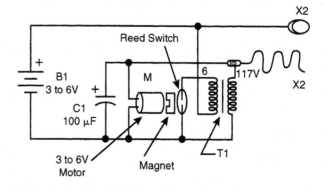

Figure 3.16.8 *Using a turning magnet to make a fast switch.*

Using a Relay as an Inverter

Figure 3.16.9 shows how a relay can be used as a vibrator to turn the current on and off across the transformer's primary coil, thus generating high voltage in the secondary coil. The characteristics of the C1 capacitor must match the relay characteristics in order to get the best performance from the circuit.

Parts List—Using a Relay as an Inverter

K1   6-volt × 50 mA relay

T1   Transformer (as in the basic project)

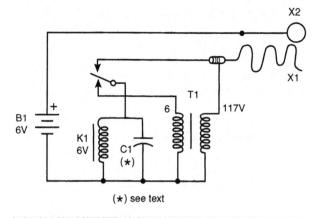

Figure 3.16.9   *Using a relay as a vibrator.*

B1   6 volts of power (four C or D cells and holder)

X1, X2   Wires as in the basic project

C1   0.01 to 0.47 µF polyester capacitor (see text)

Wires, solder, plastic box, etc.

A High-Voltage Oscillator

Figure 3.16.10 shows a power oscillator or inverter that can raise the voltage of two AA, C, or D cells to more than 100 volts, the amount necessary to cause a good shock.

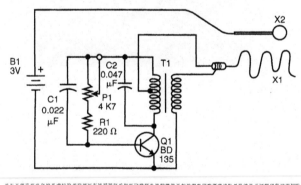

Figure 3.16.10   *A transistorized inverter.*

Notice that this circuit continues to operate even when the weaving wire remains in contact with the loop, because the circuit doesn't require changes in the current to operate.

Any small transformer with a primary coil rated to 117 VAC, a secondary coil rated from 3 to 6 volts, and current in the range of 150 to 250 mA can be used.

P1 adjusts the frequency to the best performance for the high-voltage generation.

Parts List—High-Voltage Oscillator

Q1   BD135 or equivalent medium-power *negative-positive-negative* (NPN) silicon transistor

T1   Any small transformer (see text)

P1   4.7 kΩ trimmer potentiometer

R1   220 Ω × 1/8-watt resistor (red, red, brown)

C1   0.022 µF ceramic or polyester capacitor

C2   0.047 F ceramic or polyester capacitor

B1   3 volts of power (two AA, C, or D cells with holder)

X1, X2   Wavy wire and loop, as in the basic project

Wires, plastic box, solder, etc.

The Technology Today

Based on the same operating principle but using modern technology, we find many devices that convert the low voltage from batteries and cells to high voltages. They are called inverters or AC/DC converters and they use transistors, ICs, and other modern components to drive a ferrite-core transformer or other kind of transformer in order to raise the voltages of a source.

You can find a device that uses this principle to raise the voltage of the car battery (12 V) to 117 VAC or to power domestic appliances such as small TVs and fluorescent lamps. These inverters can be used in camping or places where the AC power lines are not accessible.

Exploring the Project

To increase the performance of your device or to learn more about the circuit, do the following:

- Use two cells in series to increase the voltage in the first version.
- Experiment with transformers of different sizes and electrical specifications.
- Find on the Internet how Faraday invented the transformer.
- Make a list of appliances in your home that use transformers. Can you determine the function of the transformer in each appliance?
- Find information about security rules when working with electricity.

Project 17—Robot with Sensors

The very simple robot described here uses mechanical sensors to detect obstacles. When an object is detected, the robot inverts its movement, running backward for some seconds, and then starts again running forward but in a new direction.

Common parts and cells are used in the basic project, allowing it to be constructed even by readers who have few resources or little access to parts. It is an ideal project to be implemented at a school because the builders can put all their imagination into finding solutions for the basic building process.

One idea is to explore a competition to find the best robot (maybe the one who can escape first from a labyrinth) or even a battle where the robots pop a rubber balloon attached to the enemy.

Objectives

- Build an autonomous robot using bumper sensors.
- Program competitions with robots.
- Learn how sensors operate.
- Learn about robots and gearboxes.
- Add new sensors and circuits to the original project.

The Project

The basic idea is to build a small autonomous robot that can sense obstacles, recede when they are detected, and look for a new direction to continue the movement. As Figure 3.17.1 shows, two bumper sensors are placed in the front part of the robot. The sensors are made to look like insect antennas and act on microswitches, a very simple solution to be adopted in low-cost projects.

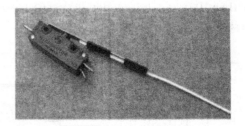

Figure 3.17.1 *The basic configuration for the robot.*

When any sensor is activated by an obstacle, a circuit formed by two monostable blocks is triggered. As Figure 3.17.2 shows, the circuits control the direction of the motors that move the robot.

The time frame in which the output of the circuit is on is different for each circuit. So when the motors are inverted, making the robot run backward, one of the motors is the first to invert its direction again. The result is that the robot spins, and when the two

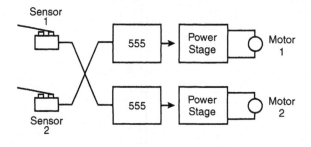

Figure 3.17.2 *Two motors are controlled by two 555 monostables.*

motors come back again to the normal forward operation, the robot runs in another direction, as shown in Figure 3.17.3.

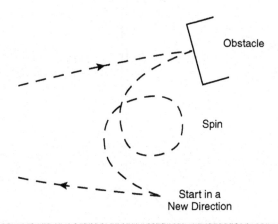

Figure 3.17.3 *The trajectory of the robot after a bump.*

The critical point for the project is simply to have a good adjustment for the times the relays are activated by the sensors. The reader must make experiments to develop the best robot performance.

The mechanical part is not critical because two gearboxes and a free wheel are placed in a small wooden or plastic chassis, as shown by Figure 3.17.4.

Gearboxes found through model dealers or in toys can be used. It is important that they are equal so the robot will advance in a straight line.

Of course, the builder is free to create different structures for the basic version and even add improvements such as other sensors, light and sound effects, and weapons (if you are building a combat robot).

The robot is powered from common cells. If your robot is lightweight, common alkaline AA cells are enough. But if your robot is heavy, you must power it from D cells or even a rechargeable 6-volt battery.

How to Build

As with any mechatronic project, this can be divided into two parts: (1) the mechanical and (2) the electronic.

Electronic Circuit

Let's start from the schematic of the electronic part shown in Figure 3.17.5.

The circuit can be powered by any 5- to 12-volt power supply. Choose according to the motor. Remember that the size of the motors and the gearbox will determine the power delivered by the robot and what it can transport.

The relays are the *dual-in-line* (DIL) type. If other types are used, the builder must make changes in the *printed circuit board* (PCB) pattern.

When mounting, it is necessary to take care with the position of polarized components such as the *integrated circuits* (ICs), diodes, transistors, electrolytic capacitors, and the power supply. The sensors are made from microswitches. Antennas, resembling the ones found on insects, can be crafted from bare wires. Equivalent replacements for the transistors can be used. Any general-purpose *negative-positive-negative* (NPN) silicon transistor can be used.

Mechanical Part

A piece of wood or plastic cut in the format shown in Figure 3.17.6 can be used as the chassis.

Gearboxes are attached to the rear and, in the frontal part, to a mobile wheel. This kind of wheel can be found in office furniture, such as chairs. Be sure that the gearboxes have the necessary reduction rate to give a good speed and torque to the robot.

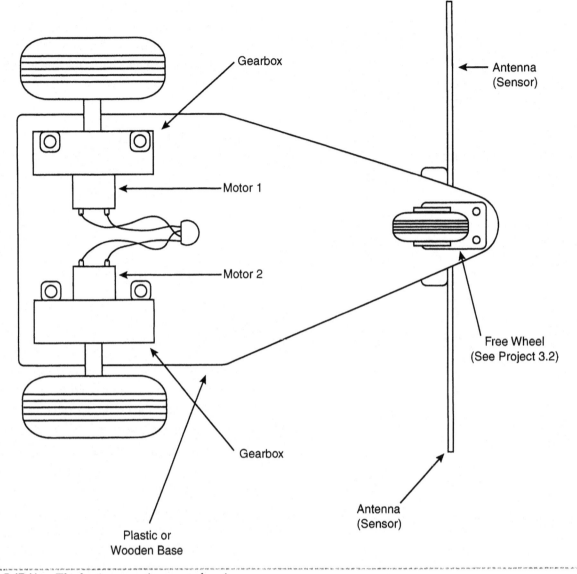

Figure 3.17.4 *The basic mounting on a chassis.*

The bumper sensors are made from microswitches, as described previously. It is important to guarantee that during any contact with an obstacle the microswitches close their contacts for a long enough time interval to trigger the circuit.

The reader is free to experiment with other configurations for the robot and even create other kinds of autonomous vehicles.

Test and Use

Verify that the sensors are okay. Place the batteries or cells in the holder, and turn on the circuit by S1.

The robot should start running forward in a straight line. If one or both motors run backward, invert their connections.

When bumping into an obstacle, the motors should invert their directions, causing the robot to back up. Adjust P1 and P2 so this movement will last only a short time. After the programmed time is exceeded, the robot will advance again.

Parts List—Robot with Sensors

IC-1, IC-2 555 IC timer

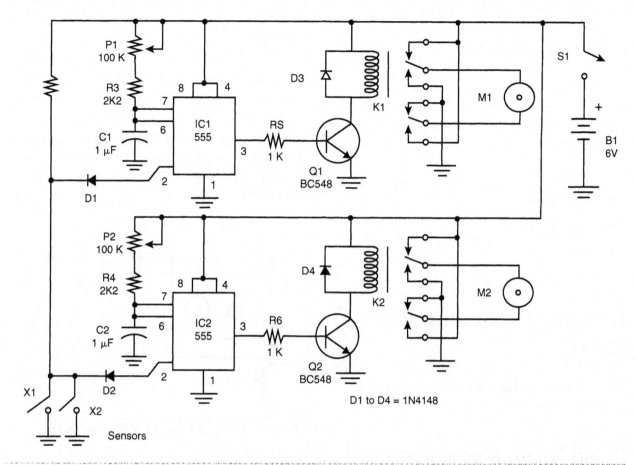

Figure 3.17.5 *Schematic for the electronic circuit of the robot.*

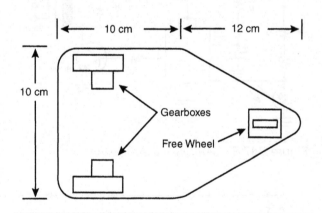

Figure 3.17.6 *The robot uses a chassis with the format and dimensions shown here.*

Q1, Q2	BC548 general-purpose NPN silicon transistor
D1 to D4	1N4148 general-purpose silicon diodes
R1, R2	22 kΩ × 1/8-watt resistors (red, red, orange)
R3, R4	2.2 kΩ × 1/8-watt resistors (red, red, red)
R5, R6	1 kΩ × 1/8-watt resistors (brown, black, red)
P1, P2	100 kΩ trimmer potentiometer
K1, K2	6-volt × 50 mA *double pole, double throw* (DPDT) relays
X1, X2	Bumper sensors (see text regarding microswitches)
S1	SPST on/off switch
B1	6 volts of power (4 AA or D cells and holder)
M1, M2	Small 6-volt DC motors
PCB, wires, solder, etc.	

Cross Themes

Because this is a simple circuit, it can be used at schools to teach crossover themes. Some elements of science courses can be linked to this project, such as the following:

- Movement
- Direction
- Control
- Time

Additional Circuits and Ideas

Many circuits can be added to the original project such as the following.

Pulse Width Modulation (PWM) control

Depending on the gearbox used in this project, the robot may advance or retract too fast to have good performance. Another problem can occur if the gearboxes are not absolutely equal; one will run faster than the other, causing the robot to advance in an unpredictable manner rather than a straight line. One way to control the speed of the motors is to add a *pulse width modulation* (PWM) control.

The circuit shown in Project 3 can be used in this project, as shown in Figure 3.17.7. The potentiometers can be used to adjust the speed of the robot with precision. See Project 3 for details about construction and operation.

Light Sensors

Figure 3.17.8 shows how a light sensor can be added to stop or invert the direction of the robot using a flashlight as a remote control transmitter. Any small LDR can be used as a sensor. For more details, see Project 20.

Timer

The circuit shown in Figure 3.17.9 can be added to the robot, giving the robot a kind of "time to think."

The circuit controlling the power supply makes the circuit stop for few seconds at regular intervals, giving the impression that the robot has stopped to think about the place it will go. R2 is chosen to give the time the robot is stopped.

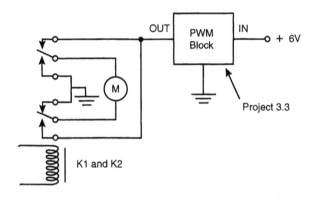

Figure 3.17.7 *Adding a PWM speed control.*

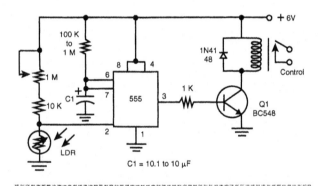

Figure 3.17.8 *Adding a light sensor.*

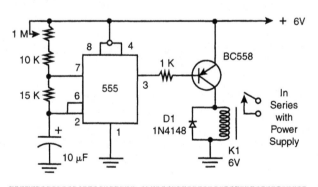

Figure 3.17.9 *A timer circuit.*

Parts List—Adding a Timer

- IC-1 555 IC timer
- Q1   BC558 or equivalent general-purpose *positive-negative-positive* (PNP) silicon transistor
- D1   1N4148 general-purpose silicon diode
- P1   100 kΩ trimmer potentiometer
- R1   2.2 kΩ × 1/8-watt resistor (red, red, red)
- R2   10 kΩ to 47 kΩ × 1/8-watt resistor (see text)
- R3   1 kΩ × 1/8-watt resistor (brown, black, red)
- C1   1 μF × 16-volt electrolytic capacitor
- C2   100 μF × 12-volt electrolytic capacitor
- K1   6-volt × 50 mA relay SPST
- PCB, wires, solder, etc.

Sound Effects

The circuit shown in Figure 3.17.10 will add a sound effect activated each time the robot hits an obstacle and the sensors are activated.

The circuit generates a sound of rising frequency when the sensors are activated and a sound of falling frequency after the time of reverse movement and until the robot stops.

The tone is adjusted by P1, and the changes in the frequency depend on R1 and R2. You can experiment with those components to find the ones with the best performance.

Parts List—Adding Sound Effects

- Q1   BC558 general-purpose PNP transistor
- Q2   BC548 general-purpose NPN transistor
- D1, D2 1N4148 general-purpose silicon diodes
- SPKR 4 or 8 Ω small loudspeaker (2.5 to 5 cm)
- R1, R2 15 kΩ to 47 kΩ × 1/8-watt resistors (see text)
- R3   1 kΩ × 1/8-watt resistor (brown, black, red)
- C1   47 μF × 12-volt electrolytic capacitor
- C2   0.047 μF ceramic or polyester capacitor

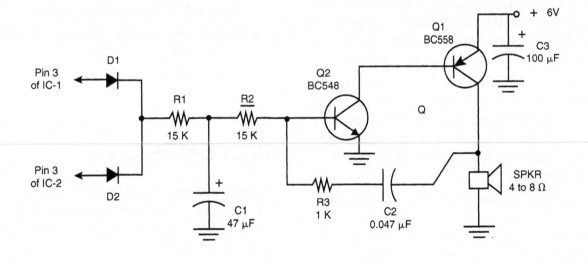

Figure 3.17.10 *A sound effect circuit.*

C3   100 μF × 12-volt electrolytic capacitor

PCB or terminal strip, wires, solder, etc.

Technology Today

Small robots with many types of sensors can be found in toy stores and bought on the Internet. Of course, the sensors can be as simple as bumper sensors or light sensors, or they can be as sophisticated as voice recognizers and sonars. The price will depend on the degree of sophistication of each type and on the number of functions it can perform.

The purpose here is to provide a project that the reader can build by him- or herself. Then the evil genius can create variations of the project by adding things not shown here.

Exploring the Project

Many ideas can be explored using this project as a starting point:

- Try to add sensors in the laterals of the robot.
- Adapt the remote control shown in Project 20 to have a robot that can move by itself but that can also be controlled by an operator.
- Design an arm (see Project 18) or add the electromagnet (see Project 6) to the robot.

Project 18—SMA Experimental Robotic Arm

Shape-memory alloys (SMAs) are materials that have the ability to return to a predetermined shape when heated. When an SMA is cold, (below a temperature called its *transformation temperature*), it has a very low-yield strength and can be deformed quite easily. However, when the material is heated above its transformation temperature, it undergoes a change in crystal structure. This change causes it to return to its original shape. If the SMA is coupled to a system that presents any resistance during this transformation, it can generate extremely large forces, providing a unique mechanism for remote actuation.

Using an SMA wire coupled to a simple mechanical arm, we can move the arm by applying an electric current. Simple movements can be obtained with only a few parts.

The aim of this project is to construct an experimental robotic arm using an SMA. Of course, the project can be upgraded to perform more functions than a simple movement, and it can even be controlled by a computer.

Objectives

- Construct a robot arm using an SMA.
- Learn how an SMA works.
- Make calculations to correctly use parts made from an SMA in mechatronic projects.
- Calculate the mechanical power delivered by a part made from an SMA.
- Determine the characteristics of an SMA.

How SMAs Work

The most common material used to make SMAs is an alloy of nickel and titanium called *nitinol*. This alloy has very good electrical and mechanical properties, long fatigue life, and high corrosion resistance.

Used in an actuator, a nitinol wire is capable of up to 5 percent strain recovery and 50,000 *pounds per square inch* (psi) restoration stress, returning to its original shape many times. It also has the resistance properties that enable it to be actuated electrically.

When heated by an electric current, it can generate enough heat to cause the phase transformation. In most cases, the transition temperature of the SMA is chosen so that room temperature is well below the transformation point of the material.

Nitinol is available in the form of wire, rod and bar stock, and thin film. For applications in mechatronics, the most common form used is the wire.

Table 3.18.1 shows some important characteristics of the SMA, according to the Robot Store catalog.

- Multiply by 0.0098 to find the force in newtons (N).
- Cycles per minute in still air at 20°C.
- LT = low temperature (70°C), HT = high temperature (90°C).

Hysteresis

Under a constant force, when the wire heats, its contraction follows the right-hand curve shown in Figure 3.18.1. When the temperature reaches the point As, the wire has started to shorten. The point of full contraction is found when the temperature reaches Af. As the wire cools, the left-hand curve is now followed, starting at the lower right and passing through Ms and Mf. Ms is the point where the wire begins to relax again, and Mf is the point where the wire is nearly fully relaxed.

SMAs can be used in applications where a short pull or push is needed. They can be considered robot muscles. Figure 3.18.2 gives an example of a typical application of an SMA wire driving a robot arm, like in our project.

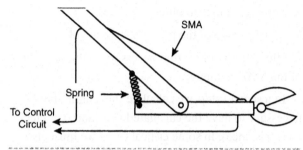

Figure 3.18.2 *A robot arm operating using an SMA.*

When the SMA is activated by a current, a force is released and the arm moves. The amount of force released can be calculated using the SMA's inherent force and the characteristics of the arm.

It is important to keep in mind the maximum force released so as not to break the wire. Therefore, the gauge of the wire must be carefully chosen according to the needs of the application (i.e., the force needed at the end of the arm).

The designer of an application using SMAs must have in mind another important characteristic: the mechanical fatigue. This characteristic relates to the number of times the SMA can contract and release, and is important because it is limited.

The circuit needed to drive the SMA must also have special characteristics. Current in excess of the amount required to release the force can break the SMA, and current less than the amount to generate the appropriate heat is not enough to reach the transition temperature; thus, the wire will not change its length. Therefore, it is necessary to see the manufacturer's specifications for a specific wire, and from this data one can design a precise current to produce the correct amount of current needed for the application.

In simple cases, knowing the voltage of the power source and the specific resistance of the SMA (ohms/m or ohms/cm), we can use simple calculations to determine the length of wire necessary for the

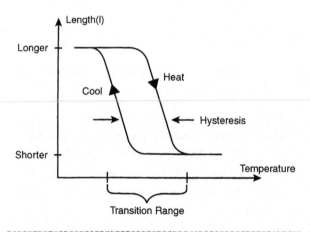

Figure 3.18.1 *The hysteresis characteristc of an SMA.*

operation current. First, we use Ohm's law to determine the total resistance required in the wire in order to enable the nominal current requirements of the SMA being used to change the shape:

$$R = V/I$$

where:

R = the resistance in ohms

V = the voltage of the power source in volts

I = the amount of current necessary to the operation of the SMA in amperes

In a second step, we calculate the length of SMA necessary to have the calculated resistance (or electric resistance). In some critical cases, we must also consider that the resistance of the SMA also changes when it is heated (the resistance increases and so the amount of current flowing is reduced):

$$L = R/P$$

where:

L = the final length, in m or cm

R = the total resistance, in ohms

P = the resistivity, in ohms/m or ohms/cm

Voltage sources commonly used to power SMA devices are cells and batteries, such as those shown in Figure 3.18.3.

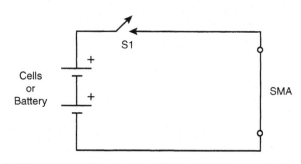

Figure 3.18.3 *Powering an SMA from cells and batteries.*

For practical purposes, we must test a current 10 to 20 percent larger than the calculated current to guarantee correct performance and compensate for voltage falls in the source. If the voltage source is higher than necessary for the length/type of SMA used in the application, the electronic potentiometer (see Project 7) or the *pulse width modulation* (PWM) control (see Project 3) can be used to adjust the power applied to the SMA.

Constant-Current Source

For the ideal operation of an SMA application, the use of a constant-current source is recommended. This kind of circuit maintains a constant current through a circuit independently from the circuit's resistance or the input voltage over a large range of values for both. Figure 3.18.4 shows a simple constant-current source using an adjustable *integrated circuit* (IC) voltage regulator such as the LM150/LM317/LM350T.

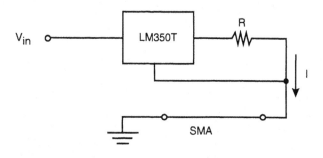

Figure 3.18.4 *Constant-current source using the LM350T.*

This IC can deliver currents up to 3 amps to a load. R depends on the desired current in the load and can be calculated by the following formula:

$$R = 1.25/I$$

where:

R = the resistance to be used in the circuit in ohms

I = the current necessary to drive the SMA in amperes

The 1.25 is the internal reference voltage of ICs like the LM150, LM350, or LM317. If another type of IC is used, its specifications sheet will give the value of the internal voltage reference that should be used in place of the 1.25 in the formula.

The voltage into the SMA depends on the resistance of the SMA or the voltage necessary to drive it.

If we know the resistance of the SMA and the current through it, we can calculate the voltage to drive it using Ohm's law:

$$V = Rsma \times I$$

where:

V = the voltage across the SMA in volts

Rsma = the resistance of the SMA in ohms

I = the current in amperes

The input voltage (Vin) must be at least 2 or 3 volts higher than the voltage of the SMA (Vsma).

The Project

The basic project consists of a simple robotic arm driven by an SMA. The robot arm makes only one movement, moving a small object up and down, as shown in Figure 3.18.5.

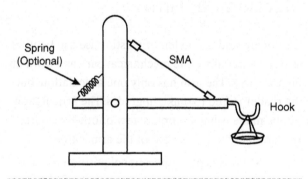

Figure 3.18.5 *Basic robotic arm.*

Of course, the evil genius is free to upgrade the project, adding other elements such as an electromagnet to pull small metallic objects or even a grip driven by solenoids. Another resource could be a second SMA muscle to create horizontal movements.

The mechanical part of this project is very simple. The electronic part is formed by a constant-current source because this is the critical point when driving the SMA. Current in excess of what is required can burn the wire, and current below what is required will not heat the wire enough to find its transition point. Also, because the current can change with alterations in the power source voltage, a mechanism to regulate the current is necessary. The current regulation circuit is made with a LM350T IC, a very easy to find and inexpensive component.

How to Build

The project is divided in two parts: (1) the electronic circuit and (2) the mechanical arm.

Electronic Circuit

The electronic circuit used to drive the SMA is shown in Figure 3.18.6. The IC must be attached to a heatsink, and P1 must be a wire-wound type, because all the SMA's current will flow across this component. P1 adjusts the transition point of the SMA.

Because a constant-current source is used, it is not necessary to calculate the length of the SMA wire in the application. This is important because many types of SMAs can be used, and you can make your arm in a large range of dimensions.

Of course, the current is limited to 3 amps, the maximum current controlled by the IC. However, for practical purposes we recommend to not go over 2 amps. The components are also calculated to a minimum current of 120 mA in the SMA.

The circuit can be mounted using a small terminal strip as a chassis for the small components, as shown in Figure 3.18.7. The electronic circuit can be housed in a small plastic or wooden box.

Parts List—Electronic Circuit for a Robotic Arm

IC-1	LM350T IC voltage regulator
D1, D2	1N5402 silicon rectifier diodes
T1	Transformer: primary coil rated to 117 VAC or according to the power supply line, secondary coil rated to 12 volts × 2 amps or according to the SMA current

Section Three The Projects

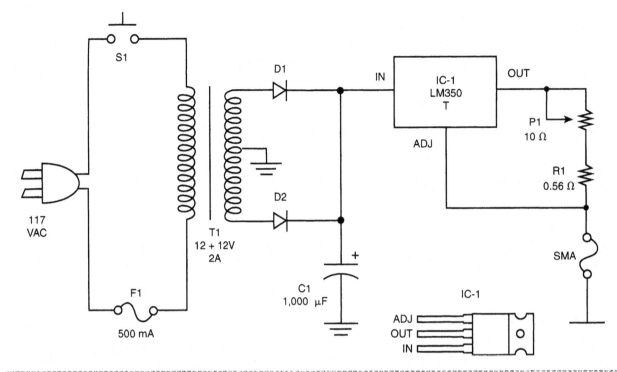

Figure 3.18.6 *Constant-current source to the SMA.*

Figure 3.18.7 *The small components can be soldered to a terminal strip.*

Mechanical Arm

Cardboard and a wooden or plastic base can be used as base material for the mechanical part, as shown in Figure 3.18.8. The arm has only one articulation, but the reader is free to make upgrades. The articulation can be made using a simple screw or other resource. Be sure that the movement of the arm is free.

The SMA is kept in operational position by two terminal strips and screws. Notice that the SMA's wires can't be soldered; therefore, they must be connected using this method.

C1	1,000 µF × 25-volt electrolytic capacitor
R1	0.56 Ω × 5-watt wire-wound resistor
P1	10 Ω wire-wound potentiometer
SMA	Any SMA with a current rate from 150 mA to 2 A
F1	500 mA fuse and holder
S1	Pushbutton (*normally open* [NO])

Wires, terminal strip, power cord, solder, terminal strip with screws, etc.

Figure 3.18.8 *The arm.*

The length of the SMA depends on the dimension of the arm. Remember that the SMA contracts about 10 percent of its length when excited. This change determines the total path of the hand in which the power will be released in order to move the weight.

Testing and Using

Great care must be taken with the SMA when testing the robotic arm in order to prevent burning it with excess current. Before plugging the power cord into the AC power line, put P1 in the position of maximum resistance, as shown in Figure 3.18.9.

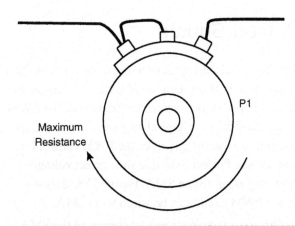

Figure 3.18.9 *Putting P1 in the position of maximum resistance (minimum current).*

After this adjustment you can plug the power cord into the AC power line and start the other adjustments. While pressing and releasing S1, turn back P1 a little at a time until the moment you observe the SMA contraction. Stop moving P1 at this moment. The circuit is now adjusted. You should not increase the current beyond this point.

Now the arm is ready for use. Carefully select the object to be moved. Don't choose something heavy because the wire has limitations to the force that it can release.

Cross Themes

It can be interesting to explore the theme of the expansion of solids when heated. Use as your example the differences between common metals and SMAs. The subject can be studied in physics as cross themes in the following topics:

- How SMAs work
- Measuring the released force
- Levers

Additional Circuits and Ideas

The following paragraphs include information on some circuits that can be used for the control of SMAs.

Constant-Current Source Using a Transistor

The circuit shown in Figure 3.18.10 can control SMAs up to 500 mA. It is a constant-current source using a transistor.

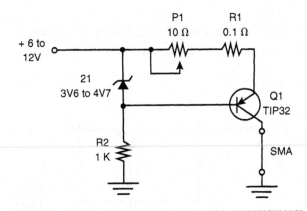

Figure 3.18.10 *A constant-current source using a transistor.*

P1 adjusts the current across the SMA. The procedure to find the correct operating point is the same as in the basic project.

The transistor must be installed on a heatsink. The input voltage can range from 6 to 12 volts, and the maximum current in the SMA is about 1 amp for this circuit.

Parts List—Constant-Current Source

- Q1 TIP32 power silicon *positive-negative-positive* (PNP) transistor
- Z1 3V6 to 4V7 × 400 mW zener diode
- R1 0.1 Ω × 1-watt wire-wound resistor
- R2 1 kΩ × 1/8-watt recistor (brown, black, red)
- P1 10 Ω wire-wound potentiometer

Printed circuit board (PCB) or terminal strip, heatsink for the transistor, wires, solder, etc.

Simple Voltage Source

In the circuit shown in Figure 3.18.11, the voltage can be adjusted according to the current needed to drive the SMA, but this is not as good as a current source. The circuit is the same as the one described in Project 7.

Again, P1 must be set initially at the minimum voltage point and adjustments are made from there.

The input voltage can be in the range of 5 to 12 volts, and the maximum output current is about 1 amp.

Parts List—Simply Voltage Source

- Q1 TIP31 power *negative-positive-negative* (NPN) silicon transistor
- P1 1 kΩ wire-wound potentiometer
- R1 220 Ω × 1-watt resistor (red, red, brown)

Terminal strip, heatsink for the transistor, wires, solder, etc.

Pulsed Source

By using a pulsed source or a source with PWM, the medium current through the SMA can be adjusted by the duty cycle of the voltage in the output. The PWM is less sensitive to voltage changes in the power supply. You can use the circuit of the PWM control for motors (see Project 3) to find the correct voltage operating point for an SMA. Figure 3.18.12 shows how a PWM circuit can be wired to an SMA.

This circuit applies a pulsed current to the SMA. The average duration of the pulses determines the current, and thus the transition point can be adjusted. The circuit can be used to drive SMAs with currents up to 1 amp.

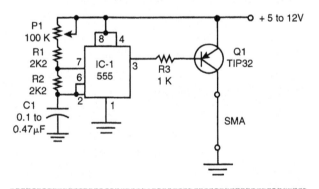

Figure 3.18.12 *Using the PWM control described in Project 3.*

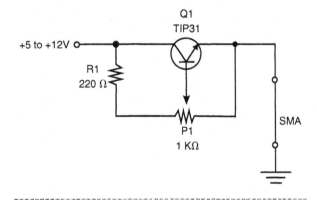

Figure 3.18.11 *A linear voltage source (electronic potentiometer) used to drive an SMA.*

Parts List—Pulsed Source

IC-1	555 IC timer
Q1	TIP32 power PNP silicon transistor
R1, R2	2.2 kΩ × 1/8-watt resistors (red, red, red)
R3	1 kΩ × 1/8-watt resistor (brown, black, red)
P1	100 kΩ linear or logarithmic potentiometer
C1	0.1 to 0.47 μF ceramic or polyester capacitor

PCB or solderless board, wires, heatsink for the transistor, solder, etc.

The Techology Today

SMAs are not used in common applications today. Because they are new and have some limitations, their use is limited to special applications such as robotics, bionics, and other experimental devices using this kind of electric muscle. Examples would include flying insects or walking robots. With the evolution of SMA technology, more and more devices will be created, including household appliances that use this kind of mechanical power source.

Exploring the Project

By knowing how SMAs work and how they can be controlled from an electronic circuit, the evil genius will be prepared to create new projects based on this mechanical power source. The following is a short list of ideas to explore:

- Design a boat using oars driven by SMAs.
- Design a walking insect robot having muscles made from SMAs.
- Create a trap or automatic door activated by an SMA.

Discovering the Characteristics of an SMA

If you have a predetermined length of SMA wire but you don't know its characteristics, it is easy to discover them using the circuit arrangement given in Figure 3.18.13. The power supply must be rated to at least 2 amps or more acording to the tested SMA.

If you have only a multimeter, you can find the current by first placing it in the A position and then, after finding the transition voltage, placing it in the B position.

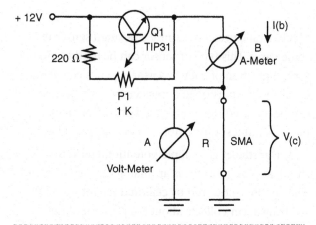

Figure 3.18.13 *Circuit used to determine the SMA's characteristics.*

Procedure

To determine the SMA's characteristics, follow these steps:

1. Adjust the output of the power (P1) supply for 0 V.
2. Open P1 slowly to increase the output voltage and simultaneously observe the SMA.
3. When the transition current is reached, the SMA will change its shape and move the test arm.
4. Increase the voltage a little more (not more than 10 percent above the previous value). Read the voltage in A.
5. Read the current in B.

6. With these values you can determine the following:

- The resistance of the SMA:

$$R = V(a)/I(b)$$

where:
R = the resistance of the SMA in ohms
V(a) = the voltage read from A
I(b) = the current read from b

- The current to drive the SMA: I(b)
- The resistivity of the SMA:

$$P = R/L$$

where:
P = the resistivity in ohms/cm
R = the resistance in ohms
L = the length in centimeters

Project 19—Position Sensor

Mechatronic projects require a position sensor in some cases. This type of sensor can be used to monitor the position of a lever, a gear, a pen, a camera, or other mobile part of an appliance. Many solutions can be adopted to monitor the position of an object, but the simplest is probably the one described here.

A transducer is a simple potentiometer (rotary or slide), and its indicator is an analog (or digital) meter. The circuit is powered by cells and the information the transducer collects about the position of the object coupled to the sensor can be sent by wires to an indicator up to hundreds of meters away.

Many applications are possible for this indicator, including an interesting game of "equilibrium" where the stress or "sobriety" of a person can be measured. Another application is to couple the circuit with a robot or mechatronic automatic device to give feedback about its operation (direction, position of the arms, and so on). The circuit can also be used for experiments in the physics and science labs, determining accurate information about the position of moving objects.

Objectives

- Build a circuit that senses the position of a remote object.
- Add feedback to robotic and mechatronic projects.
- Teach how a potentiometer works.
- Use the circuit in games and experiments.

How It Works

The idea is simple: In a circuit formed by a potentiometer, a voltage source, and a current meter, the current flowing across the meter depends on the resistance of the potentiometer. Because the resistance depends on the position of the slide, we can translate the slide position of the potentiometer into current intensity, as shown by Figure 3.19.1.

Of course, in practice some additional components must be added and some facts about the accuracy and range must be noted. First, the potentiometer has a limited range of actuation. Therefore, the distance

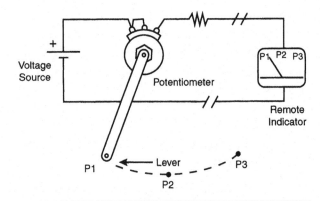

Figure 3.19.1 *Operating principle of the position indicator.*

172 Mechatronics for the Evil Genius

of the positions that can be sensed depends on the angle swept by the cursor of the potentiometer (or the length that it can slide). Figure 3.19.2 shows how rotary and slide potentiometers can be used as sensors in this project.

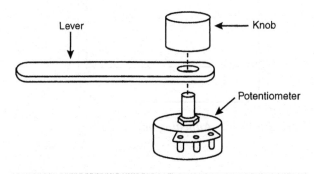

Figure 3.19.2 *How to use slide and rotary potentiometers as sensors.*

Another point to be considered is that potentiometers can present two kinds of resistance curves. In linear potentiometers, the resistance has a direct ratio to the position of the cursor. In the logarithmic (log) potentiometers, the position of the slide has a logarithmic dependency on the position, as Figure 3.19.3 shows.

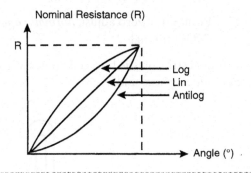

Figure 3.19.3 *Curves of common potentiometers.*

For our purposes in this project, linear potentiometers are recommended, because they present a resistance that has a direct ratio to the position of the slide. The circuit is also sensitive to the voltage of the power source, which requires an adjustment by a trimmer potentiometer and allows the operator to set the correct operation point each time he or she uses the indicator.

The circuit can operate on 3 to 6 volts of power, and the resistance of the wires between the sensor and indicator can be compensated for by the adjustment. This allows the builder to use wires as long as 100 meters.

How to Build

Figure 3.19.4 shows the complete schematic diagram for the position indicator.

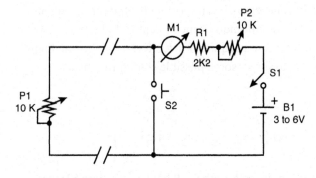

Figure 3.19.4 *Schematics for the position indicator.*

Because the circuit uses only a few small components, they can be soldered to a terminal strip and fit into a plastic box, as shown by Figure 3.19.5.

The indicator is any analog meter with a full scale between 100 μA and 1 mA. In the case of a 1 mA analog current meter, the sensor potentiometer can be reduced to 2k2 or 4k7 to get the best performance.

Figure 3.19.5 *Mounting using a terminal strip.*

When mounting, it is necessary to observe the polarity of the meter. The plus (+) must be wired to the same side of the positive line. If not, the indicator tends to move toward negative currents. If this occurs, the builder will know that he or she needs to invert the meter.

Mechanical Part

A sensor can be coupled to the potentiometer so it meets the needs of the design in many different ways. In some cases, it is important for the sensor to limit the potentiometer's movement or to have a sensor that doesn't require much effort to be moved. Figure 3.19.6 shows some possible configurations.

Testing and Using

Place the cells in the cell holder and close S1; the indicator will move. Press S2 and adjust P2 to show the full-scale indication, as shown in Figure 3.19.7.

Now you can calibrate the scale of the meter using any reference point as the base. If the wires are too long, the adjustment to zero must be done with P1 in its minimum resistance position.

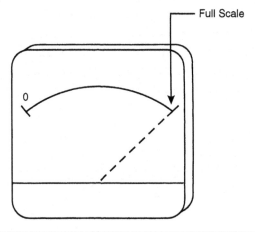

Figure 3.19.7 *Adjusting and using.*

Parts List—Mechanical Portion of Position Indicator

- P1 10 kΩ linear potentiometer (slide or rotary type; see text)
- R1 2.2 kΩ × 1/8-watt resistor (red, red, red)
- P2 10 kΩ trimmer potentiometer
- M1 100 µA to 1 mA analog meter
- B1 3 or 6 volts of power (2 or 4 AA cells)
- S1 SPST on/off switch
- S2 Pushbutton zero adjust; *normally open* (NO) optional

Terminal strip, plastic box, cell holder, wires, mechanical parts, solder, etc.

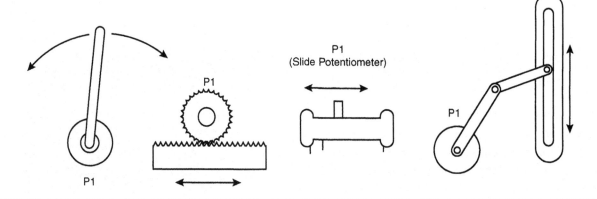

Figure 3.19.6 *Coupling the sensor.*

Cross Themes

As suggested in the introduction, the indicator can be used in a variety of science experiments. Figure 3.19.8 shows how the circuit can be coupled to a system to sense the position of a rudder in an experiment involving water flow.

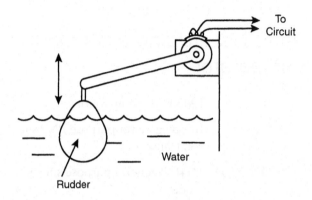

Figure 3.19.8 *Using the indicator in science experiments.*

Another experiment is very interesting because it can be made by students of human behavior. The idea is to couple the sensor to a picture that can be moved on a wall. The person must position the figure correctly according to his or her sense of equilibrium, as shown by Figure 3.19.9.

Using the indication of the circuit as a reference point, the research can discover how stress, alcohol, and other influences make individuals place the picture in a position that is incorrect. Statistics can reveal if the person tends to deviate generally to the right or left according to stress or other external influences.

Additional Circuits and Ideas

The simple circuit described in this project can be upgraded by the evil genius who is familiar with more advanced electronics. Some ideas are given here.

Using an Analog or Digital Multimeter as an Indicator

Instead of an analog indicator, you can use a common multimeter, as shown in Figure 3.19.10. The multimeter is adjusted to the lowest current scale. A table with the corresponding current and position must be made.

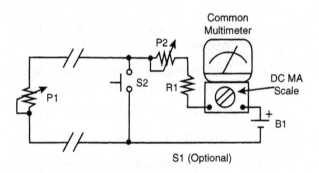

Figure 3.19.10 *Replacing the indicator with a multimeter.*

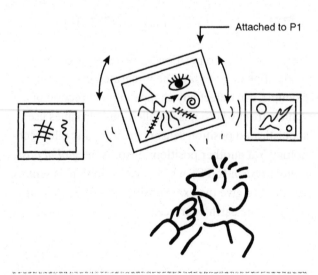

Figure 3.19.9 *An equilibrium test.*

Triggering a Relay in a Programmed Position

Another upgrade for the circuit shown in this project would be to add a comparator to trigger a relay when the sensor passes a predetermined position. This circuit uses one of the four comparators available, as shown in Figure 3.19.11.

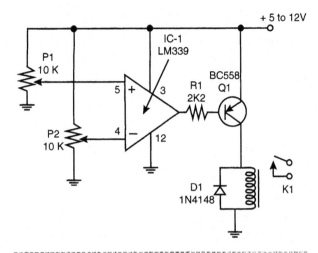

Figure 3.19.11 *Using a comparator to trigger a relay.*

The position in which the sensor will trigger the circuit is programmed by P2, and P1 is the sensor. When the voltage of P1's slide is the same as P2's slide, the comparator will trigger and drive the relay.

Parts List—Triggering a Relay in a Programmed Position

IC-1	LM339 *integrated circuit* (IC) comparator
Q1	BC558 general-purpose *positive-negative-positive* (PNP) silicon transistor
D1	1N4148 general-purpose silicon diode
P1, P2	10 kΩ linear potentiometers
R1	2.2 kΩ × 1/8-watt resistor (red, red, red)
K1	6 or 12 volts × 50 mA relay according to the power supply voltage

Printed circuit board (PCB), power supply (6 or 12 volts, wires, solder, etc.

Using a Window Comparator

The previous circuit drives the relay when the position sensor passes across a programmed position. The circuit for a window comparator, shown in Figure 3.19.12, is different. It drives the relay only when the sensor is in a narrow band of values.

The bandwidth is determined by R1, and the points that trigger the circuit are adjusted by P1 and P2. P3 is the sensor.

The circuit can be powered by 6- to 12-volt sources according to the relay. See that the cable to the sensor has three wires.

Parts List—Using a Window Comparator

IC-1	LM339 IC comparator
Q1	BC559 general-purpose PNP silicon transistor
D1, D2, D3	1N4148 general-purpose silicon diode
K1	6 to 12 volts × 50 mA relay according to the power supply voltage
R1	2.2 kΩ to 47 kΩ × 1/8-watt resistor (see text)
R2	2.2 kΩ × 1/8-watt resistor (red, red, red)
P1, P2, P3	10 kΩ linear potentiometers

PCB, three-wire cable, power supply, wires, solder, etc.

The Technology Today

Very sophisticated position sensors are found in many appliances today. The integrated gyroscope, for example, can detect any change in the direction of a vehicle from where it is installed, and airplanes use sophisticated position sensors in automatic pilot technology. Yet another position sensor is the *global positioning system* (GPS), which uses radio signals sent by satellites to give the exact position of a person at any point on the earth's surface.

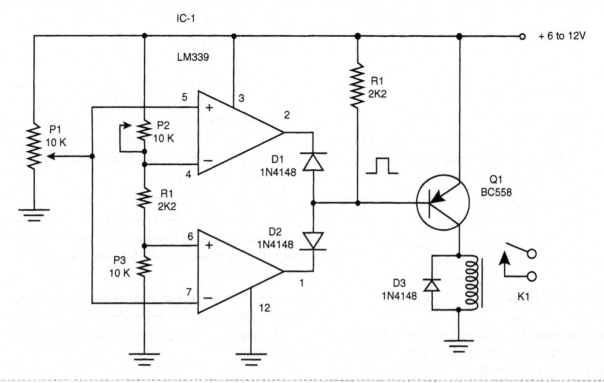

Figure 3.19.12 *Using a window comparator.*

Ideas to Explore

Ideas for experiments and improvements for this project include the following:

- Use two potentiometers at right angles to sense an object in a plane by giving the objects coordinates on the X and Y axes.

- Design an analog-to-digital conversion device to send the positions of an object to a computer.

- Use the same principles described in the analog computer (see Project 11) to make calculations regarding the positions of two objects sensed by potentiometers.

Project 20—Light-Beam Remote Control

The remote control described here can be used with many of the projects in this book. For instance, you can start your mechatronic elevator or ionic motor by simply pushing a button. You can also trigger your electronic canon with a light beam or control a mechatronic airboat or robot arm by using a remote control. You can also use the remote control as a detector, placing it in automated devices to detect when a light beam is cut or when light falls onto the sensor.

The circuit is very simple as you have to mount only the receiver. The transmitter is simply a flashlight, and the receiver has no critical parts or complicated adjustments. Of course, the range is limited to the light of the flashlight and the focusing system in the receiver, but in good conditions you can control a model placed at distances of up to 10 meters. Additional applications for the project will be suggested throughout this project.

Project 20 – Light-Beam Remote Control

Objectives

- Control models and automatic devices using a light beam.
- Study how photosensors work.
- Design projects that are sensitive to light or dark such as traps, intrusion detectors, or other automated devices.
- Apply the circuit for these automated devices to experiments in sciences.

The Sensor

The sensor used in this project is a *light-dependent resistor* (LDR). This device is also called a *cadmium sulfide* (CdS) cell from the material used in its fabrication.

The LDRs, such as the ones shown in Figure 3.20.1, present a very high electric resistance when in the dark. But when illuminated, their resistance of many megohms falls to hundreds or tens of ohms, leaving the current to flow across them.

They are very sensitive at detecting amounts of light that even the human eye can't detect in normal conditions. Although they are sensitive, they are not as fast as other electronic detectors such as photodiodes and phototransistors. For our purposes, they are ideal because they are inexpensive, easy to find, and easy to use.

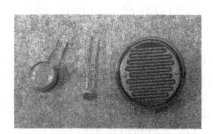

Figure 3.20.1   *Common LDRs.*

The Project

In our basic light beam project, the circuit triggers a relay when a light beam is focused on the sensor (LDR). Upgraded versions will time the action of the relay or give a bistable action. The version chosen by the evil genius will depend on the application he or she has in mind.

The operating principle is very easy to understand: When the sensor detects the light, its resistance falls, and the current can flow across the base of a transistor. This current is amplified, driving the relay that closes its contacts.

P1 is adjusted to put the transistor on the threshold of conduction with the ambient light. Therefore, any amount of light falling onto the sensor triggers the circuit.

How to Build

Figure 3.20.2 shows the schematic for the basic version of the remote control using a light beam.

The relay is chosen according to the power supply voltage. The types of relays with winding currents of up to 50 mA or more are recommended. A simple mounting can be made using a solderless board, as shown in Figure 3.20.3.

Because the circuit draws relatively low current, it can be powered by the same supply used in the application where the circuit will be installed. The relay in

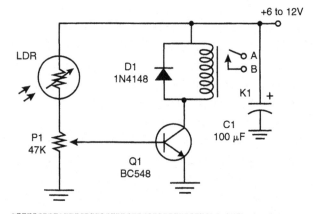

Figure 3.20.2   *The schematic for the remote control.*

Figure 3.20.3 *The circuit can be mounted using a solderless board.*

the photo is a universal type with dual inline terminals. If you use another type of relay, you must be sure about the identification of the terminals for the coil and the contacts.

The LDR can be of any size, but it must be round and installed inside a small cardboard tube or other opaque material. A convergent lens can be added to get directivity and increase the sensitivity of the circuit, as shown in Figure 3.20.4.

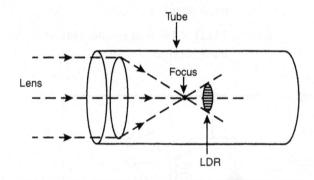

Figure 3.20.4 *Adding a lens.*

The power supply depends on the voltage rate of the relay's coil. For instance, if you use a 6 V × 50 mA relay, the power supply can be formed by four AA, C, or D cells. It is helpful to review the specifications so that you use the same supply for the circuit and the rest of the project.

Testing and Using

It is very easy to test your remote control. Power the circuit on and close P1 (put it in the position of minimum resistance). Then point the LDR toward a dark place.

Now, opening P1 slowly, you will hear when the contacts close. Open P1 a little to open the contacts again. Then, focusing the flashlight on the LDR, you will hear when the relay closes the contacts. If you can't hear your relay, a *light-emitting diode* (LED) as an indicator can be wired to the circuit, as shown in Figure 3.20.5.

When using the LED, adjust P1 to put the relay near the triggering point. Then the flashlight can trigger the circuit. Remember that the relay will be on only during the time the sensor receives the light.

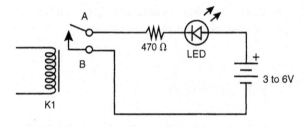

Figure 3.20.5 *Using an LED to adjust the sensitivity.*

Parts List—Using an LED to Adjust Sensitivity

Q1 BC548 general-purpose *negative-positive-negative* (NPN) silicon transistor

D1 1N4148 general-purpose silicon diode

LDR Any LDR or CdS cell (see text)

P1 47 kΩ trimmer potentiometer

C1 100 μF × 12-volt electrolytic capacitor

K1 6- or 12-volt × 50 mA relay

Printed circuit board (PCB) or solderless board, wires, solder, power supply, etc.

Cross Themes

A light-beam remote control can be used as an auxiliary device to implement or trigger experiments in many devices. The control can be used to activate devices remotely, including traps, fans, lamps, and more. A few suggestions follow:

- Close a trap made to catch animals.
- Start dangerous experiments remotely.
- Turn on and off devices in experiments from a safe distance.

Additional Circuits and Ideas

For more fun with light beams, a number of different circuits and ideas should be considered.

Timed Version

A simple timed version using the 555 *integrated circuit* (IC) is shown in Figure 3.20.6. In this circuit, P1 adjusts the sensitivity according to the light source and the ambient illumination. P2 adjusts the time the relay is on after being triggered. The time on can be calculated by the following formula:

$$t = 1.1 \times R2 \times C2$$

where:

t = the time in seconds (s)

R = the adjustment of P2 and R2 in ohms (Ω)

$C2$ = the capacitance in farads (F)

The maximum values for R2 and C are 1 MΩ and 1,000 µF.

Parts List—Timed Version

IC-1	555 IC timer
Q1	BC548 or equivalent general-purpose NPN silicon transistor
D1	1N4148 general-purpose silicon diode
P1, P2	1 MΩ trimmer potentiometers
R1	47 kΩ × 1/8-watt resistor (yellow, violet, orange)
R2	10 kΩ × 1/8-watt resistor (brown, black, orange)

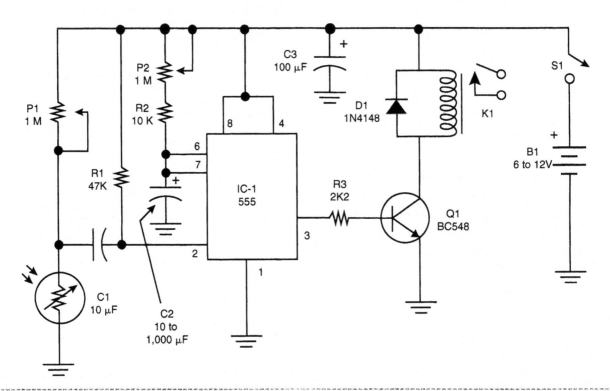

Figure 3.20.6 *Timed version using the 555.*

R3	2.2 kΩ × 1/8-watt resistor (red, red, red)
C1	10 μF × 12-volt electrolytic capacitor
C2	10 to 1,000 μF × 12-volt electrolytic capacitor
C3	1,000 μF × 12-volt electrolytic capacitor
LDR	Any common LDR (see project)
K1	6- or 12-volt relay, 50 mA coil (according to the power supply)
B1	6- or 12-volt cells or power supply (according to the relay)

PCB or solderless board, power supply, wires, solder, etc.

Bistable Version

Figure 3.20.7 shows a bistable version based on a configuration explored in previous projects using the D-type *complementary metal oxide semiconductor* (CMOS) flip-flop 4013.

P1 adjusts the sensitivity and the relay must be rated according to the power supply voltage. Types rated from 5 to 12 volts can be used.

The circuit can be mounted on a PCB or on a solderless breadboard. Take care with the position of polarized components such as the ICs, electrolytic capacitors, transistor, and power supply.

Parts List—Bistable Version

IC-1	555 IC timer
IC-2	4013 CMOS D-type flip-flop IC
Q1	BD548 general-purpose NPN silicon transistor
D1	1N4148 general-purpose silicon diode
K1	5- to 12-volt relay (according to the power supply voltage)
LDR	Any round, common LDR
R1	10 kΩ × 1/8-watt resistor (brown, black, orange)
R2, R3	100 kΩ × 1/8-watt resistors (brown, black, yellow)
R4	2.2 kΩ × 1/8-watt resistor (red, red, red)
P1	1 MΩ linear or logarithmic potentiometer (sensitivity)
C1, C2	0.1 μF ceramic or polyester capacitors

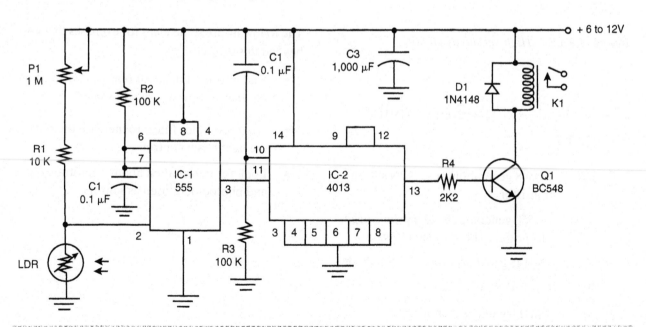

Figure 3.20.7 *Bistable version using the 4013 CMOS IC.*

C3 1,000 µF × 12-volt electrolytic capacitor

PCB, wires, solder, power supply, etc.

High-Sensitivity Version

Very high sensitivity can be achieved with the circuit shown in Figure 3.20.8. A second transistor is added to increase the sensitivity of the circuit.

When the sensor is installed in a tube with a convergent lens, the circuit can detect light sources up to 20 meters away and even farther if in the dark (if little ambient light exists).

P1 adjusts the sensitivity, and the operating principle is the same as in the basic project. The circuit will remain on only during the time the light is shining on the sensor. The same input stage can be adapted from the other projects (mono and bistable versions).

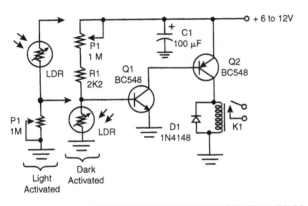

Figure 3.20.8 *High-sensitivity circuit.*

Parts List—High-Sensitivity Version

Q1 BC548 general-purpose NPN silicon transistor

Q2 BC558 general-purpose *positive-negative-positive* (PNP) silicon transistor

D1 1N4148 general-purpose silicon diode

K1 6-volt V × 50 mA relay

P1 1 MΩ trimmer potentiometer

R1 2.2 kΩ × 1/8-watt resistor (red, red, red)

LDR Any round LDR or CdS cell (see text)

C1 100 µF × 12-volt electrolytic capacitor

PCB or terminal strip, wires, solder, etc.

The Technology Today

Today many of the remote controls for appliances (domestic and industrial) operate on *infrared* (IR) or *radio frequency* (RF) signals. These remote controls use an advanced technology that utilizes the signals so that each control manipulates only the appropriate and corresponding receiver.

Microprocessors and other resources are used to give millions of different codes, thus making the controls absolutely safe. Other controls also have the capability to command many functions in the receiver.

Of course, a simple remote control, such as the one in this project, has the advantage that it is not critical, it can be mounted by anyone, and it uses only a few inexpensive parts.

Ideas to Explore

The following are some ideas for exercises that go beyond this project:

- Find information about remote controls that use IR and Phase Locked Loops.

- Try to use color filters to increase the range of the light beam control.

- Look for information about phototransistors and photodiodes used as sensors.

Project 21—Mechatronic Airboat

This is a very simple project based on the same principles as the race car (see Project 1). A blade propeller is used to move a boat, such as the airboats used in the Everglades and other swampy or wetlands areas.

Because no parts are critical and all parts are inexpensive and easy to find, this is an ideal project to adopt, with easily identifiable middle-school cross themes. Students of Colegio Mater Amabilis in Brazil built this ship and participated in a competition, as shown in Figure 3.21.1.

Some upgrades are suggested to get even more from the mechatronic airboat. You can add a remote control, a timer, and special effects such as lights and sounds.

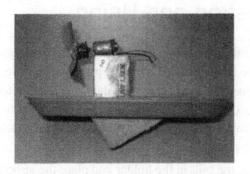

Figure 3.21.1 *Mechatronic ship built by young students.*

Objectives

- Build a small experimental airboat.
- Organize airboat competitions.
- Study the Newton's laws as a cross theme.
- Study the movement of ships in the water.

The Project

The basic project consists of a small plastic or styrofoam tray (such as the ones used to hold produce and meat in supermarkets) in which a motor and cells are installed. The motor rotates a fan that acts as a propeller forcing the air in one direction and therefore moving the ship in the opposite direction.

A few factors must be considered when assembling the airboat. The first factor is the correct positioning of all the parts on the tray in order to maintain the boat's equilibrium. As shown in Figure 3.21.2, if the heavy parts, such as the cells, are positioned incorrectly, the boat will sink.

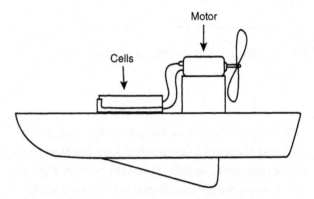

Figure 3.21.2 *Heavy parts must be correctly positioned on the tray.*

The design of the boat's hull should also be considered. For example, the height of the tray is important in order to guarantee that it will float. How much of the tray's height that will be under water, compared to the amount that remains above water (the freeboard), depends on the weight of the tray's parts. Another important factor is the keel, which helps the boat navigate in a straight line.

Finally, we have to consider the efficiency of the fan. The more efficient the fan is, the higher the speed of the airboat will be. If you intend to enter your airboat in a competition, this is a point you can't forget during the design stage.

The small DC motor can be powered from two or four cells according to the type of motor. In a competition, it is important to have all the ships use the same power supply.

How to Build

Figure 3.21.3 shows the electric circuit for the airboat. A switch to turn the motor on and off is optional. You can start the motor by simply inserting the cells in the holder.

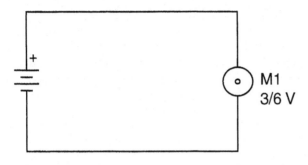

Figure 3.21.3 *The electric circuit used in the airboat.*

If you are a beginning evil genius, take care when soldering the wires because a bad joint can stop your motor right in the middle of the competition. Figure 3.21.4 shows the airboat. Notice that the cells are in the front of the tray (the bow of the boat) to balance the weight of the motor and its support.

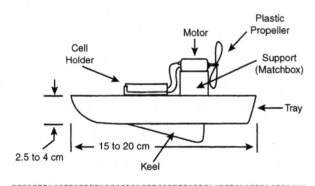

Figure 3.21.4 *The airboat.*

The motor is placed in the rear, or the stern, of the boat. The support for the motor can be a small plastic box where the motor is glued. You can use a small piece of light wood or even cardboard. The keel is made from a piece of styrofoam or other material and has the form shown in Figure 3.21.4.

The propeller is made from plastic or wood with two to four blades. These can be found in stores that sell supplies for models. The diameter of the prop should be determined by the weight and size of the boat. The maximum recommended diameter is 10 centimeters.

Test the balance of the ship in a small basin until you find the best positioning for the parts.

Parts List—Electric Circuit

B1 3 or 6 volts of power (2 or 4 AA cells with holder)

M1 3- or 6-volt DC small motor

Wires, solder, etc.

Testing and Using

The initial tests should be made during construction to see if the parts are correctly placed in the tray. Because the correct position should have been found by this point, you can now test the propeller system in a larger body of water such as a small lake, tank, or other place.

Put the cells in the holder and place the airboat in the water. If it doesn't move in a straight line, change the position of the cell holder or the propeller. Now you only have to organize a competition or upgrade your project.

Exploring the Project

Many improvements can be added to the original design. Depending on what you intend to add, the airboat may need to be enlarged to support the weight of the additional elements. Some ideas are given here:

- **Pulse width modulation (PWM) or linear control:** The PWM and electronic potentiometer (refer to Projects 3 and 7) can be used in versions that use 6-volt motors. You can replace the potentiometers with trimmer potentiometers and adjust the speed of the propeller according to your needs.

- **Adding a rudder**: A small rudder can be added, as shown in Figure 3.21.5. It can be used to direct the movement of the airboat.

- **Creating a competition**: An airboat competition can be organized in many ways. A water tank can be built using a large piece of plastic or tarp, such as the kind used to protect objects from rain or weather. It can be placed in a box made from wood as shown in Figure 3.21.6.

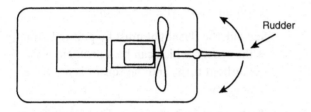

Figure 3.21.5 *Adding a rudder.*

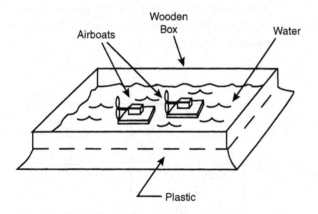

Figure 3.21.6 *A simple tank in which to run airboat competitions.*

This tank can be filled with 10 to 15 centimeters of water, enough to simply allow the airboats to move freely. You can make lanes using wooden slats, as shown in Figure 3.21.7. The reader is free to create the appropriated rules for the competition.

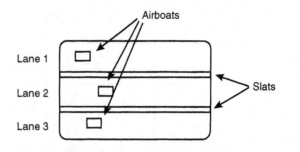

Figure 3.21.7 *Lanes made from slats.*

Cross Themes

The characteristics of floating objects, Newton's law of propulsion, and the speed and direction of an object in water are all themes studied at some point in the science and physics curricula. Teachers can use this airboat project to perform experiments and illustrate the theories being taught. The main subjects that can be explored include the following:

- Newton's law
- Floating
- Balance
- Speed

Additional Circuits and Ideas

Many improvements can be added to this project to get higher performance from the airboat. Some of them include the following:

- **Two motors and a differential remote control**: Figure 3.21.8 shows how two motors can be used to give a directional control to the airboat. The speed of the motorc can be controlled by a two-channel remote control.

  When the two motors are running at the same speed, the ship advances in a straight line. If one motor runs faster than the other, the airboat turns in the direction of the motor that is running slowly.

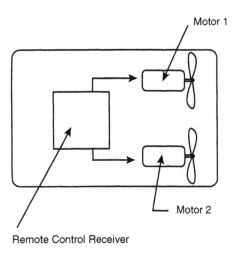

Figure 3.21.8 *Using two motors and a remote control system.*

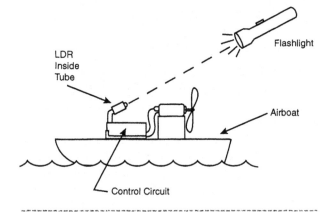

Figure 3.21.10 *Using a light-beam remote control.*

- **Remote control with a rudder**: Figure 3.21.9 shows how a solenoid can be used to move a rudder installed on the airboat. The solenoid can be energized by a one-channel remote control. Using the remote control, the ship will turn right and left, allowing the evil genius to navigate the airboat in any direction or through a course.

- **Control by using a light beam**: Figure 3.21.10 shows a simple remote control using a flashlight. More details about this control can be found in Project 20.

This simple control can be used to trigger a solenoid coupled to the rudder. By pulsing the light, the evil genius can change the direction of the airboat.

When using this control, two points must be considered. First, the sensor must be installed in a tube to receive only the light from the flashlight. Daylight cannot reach the sensor or it will interfere with the circuit.

Second, the flashlight must be pointed directly at the sensor, which can be a problem when the airboat turns and the sensor is no longer facing you.

- **Add a sound-effect circuit**: The circuit shown in Figure 3.21.11 produces a sound like that of a motor boat. The circuit can be powered from 3- to 6-volt power supplies. The speaker is a small one, with a diameter of only 2.5 to 5 centimeters (1 to 2 inches).

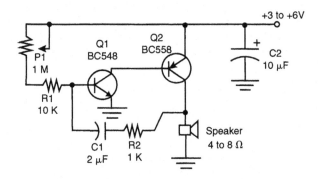

Figure 3.21.11 *Motor boat sound-effect circuit.*

It is important to place the circuit in a position where no water can affect its parts, mainly the loudspeaker's cone, which is normally made from cardboard. P1 adjusts the sound, so find the position that sounds like a motor boat. Figure 3.21.12 shows how to mount this simple circuit using a terminal strip as a chassis.

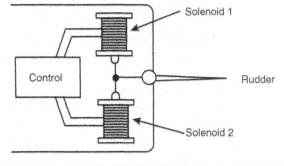

Figure 3.21.9 *A simple one-channel remote control moving a rudder.*

186   Mechatronics for the Evil Genius

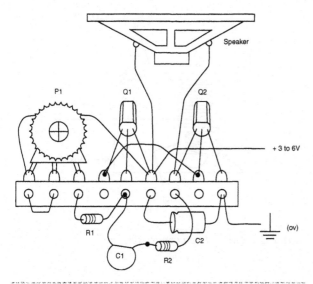

Figure 3.21.12 *Motor boat sound effect built using a terminal strip.*

Parts List—Motor Boat Sound Effect

- Q1 BC548 or equivalent general-purpose silicon *negative-positive-negative* (NPN) transistor
- Q2 BC558 or equivalent general-purpose silicon *positive-negative-positive* (PNP) transistor
- R1 10 kΩ × 1/8-watt resistor (brown, black, orange)
- R2 1 kΩ × 1/8-watt resistor (brown, black, red)
- P1 1 MΩ trimmer potentiometer
- C1 1 μF polyester capacitor
- C2 10 μF × 12-volt electrolytic capacitor
- SPKR Small 4 or 8 Ω loudspeakers (2.5 to 5 cm)

Terminal strip or *printed circuit board* (PCB), solder, wires, plastic box, etc.

The Technology Today

Propellers are used in the design of many types of vehicles. In the desert and on the ice, propellers help transport cars, and in the wetlands propellers are used for boats. Of course, this type of propulsion has some disadvantages because the power is transferred directly to wheels.

Ideas to Explore

- Make a turbine by inserting the propeller into a tube, creating a jet ship.
- Experiment with several types of propellers and different numbers of blades to find the best performance for your boat.
- Design a propeller using a small fan such as the ones used in computers to lower the temperature of microprocessors.

Project 22—Coin Tosser

Solenoids are important components in mechatronic projects. The reader who has an unlimited imagination can create infinite projects using solenoids as the basic element.

This book explores many applications for solenoids such as the canon, the galvanometer, the magic pendulum, and projects performed with the laser Lissajous. Here is another application for solenoids: a simple project that launches a coin into the air in a completely random manner without tricks or manipulations by the operator.

Objectives

- Build a project that uses a solenoid as its basis.
- Learn more about solenoids.
- Use a coin tosser to get simple yes/no decisions.

How It Works

The basic idea is very simple: When energized, the magnetic field produced by a solenoid pulls a metallic core through the center of the coil. This fast movement launches a small coin into the air, as shown in Figure 3.22.1.

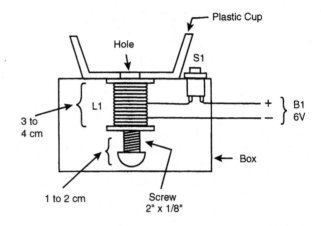

Figure 3.22.2   *The coin tosser.*

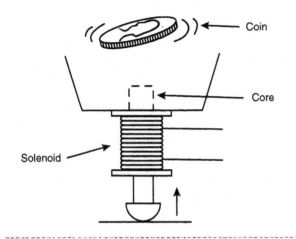

Figure 3.22.1   *The basic idea of a coin tosser.*

The critical issue is to build the solenoid strong enough to launch a coin when powered from C or D cells. Because enameled wires come in many gauges and the magnetic field depends on the number of turns and current across the coil, a large range of construction possibilities must be examined by the reader.

In the text we will give the basic version, which has been tested by the author and provided good results. But depending on the material used, mainly the weight of the core and dimensions of the solenoid, it will be necessary to test your project.

How to Build

Figure 3.22.2 shows the basic version of the simple coin tosser. The circuit can be powered by four C or D cells because the current required is too high to be sourced from AA cells (depending on the resistance of the wire used in the coil).

The coil is wound around a cardboard or plastic bobbin with the dimensions shown in the figure. The enameled copper wire can be 28- to 30-gauge wire, and the number of turns should be in the range of 200 to 500, according to the dimensions of the bobbin and the gauge of the wire. The core is a 2- × 1/8-inch screw that must have free movement when sliding inside the coil.

Notice that the core must be partially inside the bobbin when the circuit is off. You will need to experiment with the position so that when the coil is energized the core will be pulled inside.

Testing and Using

Adjust the position of the core inside the coil, as recommended previously. Press S1 and verify if the core is pulled quickly toward the inside of the coil. If not, change the position of the core and/or change the number of turns of the coil. If the operation is okay, put a small coin in the tray and see if it is launched when S1 is pressed.

Parts List—Coin Tosser

L1   Coil (see text)

S1   Pushbutton (*normally open* [NO])

B1   6 volts of power (four C or D cells)

Wires, cell holder, core for the coil, solder, etc.

Cross Themes

A coin tosser can be used in experiments involving the random generation of probable situations. In science, a coin tosser that can't be manipulated or tricked by the operator is very useful. Other applications with cross themes in the sciences are suggested here:

- Use the project to generate random numbers in statistics research.
- Use the project to launch dies.

Additional Circuits and Ideas

Other changes in the circuit can be made by the reader.

Capacitive Discharge Circuit

The basic idea of the electronic canon (refer to Project 9) to increase the power applied to a solenoid can also be applied to this project. The circuit shown in Figure 3.22.3 charges a high-value capacitor with about 36 volts. When pressing S1, the *silicon controlled rectifier* (SCR) turns on and allows the capacitor to discharge across the coil.

The energy produced by the capacitor is very high, creating a strong pull for the coil. As in the electronic canon, the pull is enough to launch the coin a good distance. The reader can use capacitors in the range of 4,700 to 22,000 µF. The pull will also depend on the coil.

When mounting, be careful with the polarized components such as the diodes, SCRs, and electrolytic capacitors. If any of them becomes inverted, the circuit will not operate as expected.

Coils with 100 to 200 tuns of 28 or 30 AWG wire can be used. The initial current released by those coils can reach 40 amps or more. Test to find the best coil and the best capacitor to launch the coin with the desired force.

The operation is the same as in the canon project. First charge the capacitor by pressing S1, and then after releasing this switch, press S2 to trigger the SCR. The *light-emitting diode* (LED) will indicate when the capacitor is fully charged.

Parts List—Capacitive Discharge Circuit

SCR   TIC106 or equivalent SCR

D1   1N4004 silicon rectifier diode

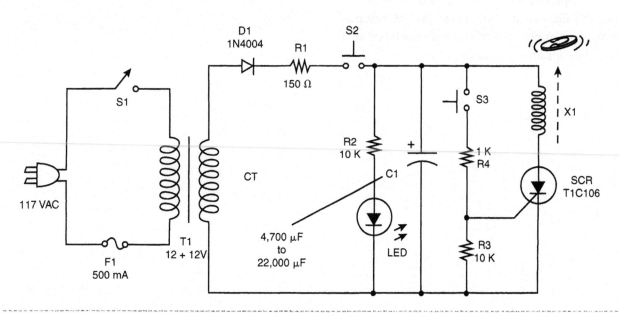

Figure 3.22.3 *The capacitive discharge circuit for a coin tosser.*

LED	Common LED (any color)
T1	Transformer: primary coil rated according to the AC power line, and secondary coil rated to 12 + 12 volts (*center tap* [CT]), 250 to 500 mA
R1	150 Ω × 1-watt resistor
R2, R3	10 kΩ × 1/8-watt resistors (brown, black, orange)
R4	1 kΩ × 1/8-watt resistor (brown, black, red)
C1	4,700 to 22,000 μF × 40-volt electrolytic capacitor
X1	Coil (see text)
S1	On/off switch
S2, S3	Pushbutton (NO)
F1	500 mA fuse and holder

Power cord, *printed circuit board* (PCB) or terminal strip (see the mounting of the canon in Project 9), wires, solder, etc.

Catapult

Another way to launch coins is by using a catapult, as shown in Figure 3.22.4.

The number of turns for the coil and the mechanical system to launch the coin must be designed by the reader. Experiment to match the force of the solenoid with the weight of the coin to find the best performance. A super powerful catapult can be created using the capacitive discharge circuit.

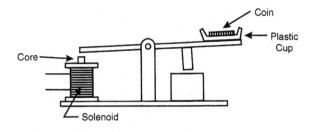

Figure 3.22.4 *Using a catapult.*

The Technology Today

Pulling objects using the magnetic force of a solenoid is a function found in many common appliances. We have discussed these appliances when working on other projects in this book that use solenoids, such as the canon and the magic motion machine.

Exploring the Project

Many alterations can be made to this project so that it can be used in other applications. Try the following:

- Use the solenoid to move a mobile figure.
- Add a timer to launch the coin automatically.
- Add a sensor to trigger the circuit with the touch of your fingers.

Project 23—The Challenge of the Mechatronic Timer

This project provides a simple but interesting idea for a competition involving mechatronics and the skills of the evil geniuses who like this kind of challenge. The basic idea is to construct a timer using only mechanical parts that turn on a lamp after a predetermined time interval. The winner might be the person who has the project with the highest accuracy and the most imaginative solution.

The circuit is very simple, and the same configuration can be used in many other projects that could be ideal for a competition organized by students or other groups.

Objectives

- Build a mechanical timer that triggers an electronic circuit for a precise time period.
- Find solutions for the highest accuracy.
- Study the movements of mechanical parts.
- Participate in a competition involving skills in mechatronics.
- Use the circuit to study cross themes of middle school and high school curriculum.
- Add the basic configuration to create other automatic devices.

How It Works

This project can be divided into two parts: the mechanical part and the electronic part.

Mechanics

The basic idea is to trigger a circuit using a magnet with a reed switch. The mechanical parts must move in a manner that passes the magnet near the reed switch, thus triggering the circuit for a preprogrammed time period, as suggested by Figure 3.23.1.

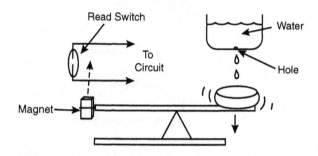

Figure 3.23.1 *The basic idea of the mechatronic timer.*

The challenge is to design a mechanical system that places the magnet near the reed switch only during the preprogrammed time. Many solutions can be adapted such as the ones shown in Figure 3.23.2.

Figure 3.23.2 (a) shows a simple idea using hydrotronics, or the use of hydraulics combined with mechanics and electronics. The water flows slowly from one reservoir to another, causing a float on which a magnet rests to rise. When the float reaches the reed switch, the circuit is triggered. Figure 3.23.2 (b) shows a different solution, one using a lever. In this case, the increasing weight in one side causes the other side to rise, thus aligning the magnet with the reed switch after the programmed time period. Figure 3.23.2 (c) shows yet another interesting solution using a stepped rotary motion system driven by a pendulum.

Of course, the imagination of the reader can find other solutions such as antique clock mechanisms driven by springs and so on.

Electronics

The electronic portion of the circuit is very simple. The reed switch can't control large loads because in many parts the maximum current is limited to

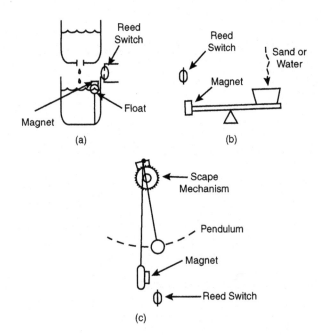

Figure 3.23.2 *Some suggested solutions for the challenge.*

200 mA or even less. Thus, we used the reed switch to trigger a *silicon controlled rectifier* (SCR), as shown in Figure 3.23.3.

This circuit latches when triggered and powers any load wired to its anode. You can power a 6-volt lamp, a buzzer alert, an electronic siren, a relay, or even a small motor driving another appliance. To reset the circuit, turn off the switch (SW1) that controls the power source and turn it on again after removing the magnet from the reed switch.

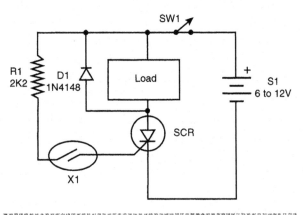

Figure 3.23.3 *The electronic circuit for the challenge.*

It is important to remember that, when the SCR conducts, a voltage of about 2 volts is observed between the anode and cathode. Therefore, when 6-volt loads are driven, it is important to compensate for this voltage fall by adding 2 or 3 volts to the supply. A 9-volt supply is recommended in this case.

How to Build

The schematic for the electric timer is shown in Figure 3.23.3. No critical parts are used, and the evil genius is free to make changes in the original project. Because it is a very simple configuration, the reader can mount it using a terminal strip to hold the components, and Figure 3.23.4 shows this mounting. When mounting, take care positioning the polarized components such as the SRCs, diodes, and power supply.

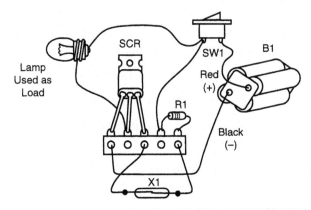

Figure 3.23.4 *A terminal strip is used to hold the components.*

The wire to the sensor can be long, up to 10 meters if necessary, and does not need to be shielded or insulated. The diode is used only if the load is inductive such as with relays, motors, and solenoids. Lamps and electronic circuits don't require the diode.

Testing and Using

Testing is very simple. Insert the cells in the holder and turn on the circuit acting on SW1. The load will remain off.

Now place the magnet near the sensor. The SCR will be triggered, and the load will turn on. To reset the circuit, remove the magnet and place it far away from the sensor. Then open and close SW1 again. If everything functions as described, you can use the circuit in your challenge.

Parts List—Electric Timer

- SCR TIC106, C106, or MCR106 common SCR
- D1   1N4148 (only for inductive loads)
- R1   2.2 kΩ × 1/8-watt resistor (red, red, red)
- SW1 On/off SPST switch
- X1   Reed switch
- Load Any 6- or 12-volt load (according to the power supply voltage)
- B1   4 to 8 AA cells or power supply from the AC power line (6 to 12 volts)

Terminal strip, solder, wires, magnet, etc.

Cross Themes

The challenge proposed by this project is perfect to use in discussions of principles from physics, mechanics, and mathematics. The teacher can associate the challenge with the following themes:

- Pendulum
- Time measurements
- Levers
- Hydraulics
- Gears

Through a timer competition, students can find the best solutions based on knowledge acquired in their regular curriculum.

Additional Circuits and Ideas

The following ideas are for additional circuits based on the mechatronic timer.

An Audio Oscillator

Figure 3.23.5 shows a simple audio oscillator driven by the SCR when the sensor is closed. The circuit can be easily mounted using a terminal strip as a chassis. Observe the position of the polarized components as the transistors and power supply lines.

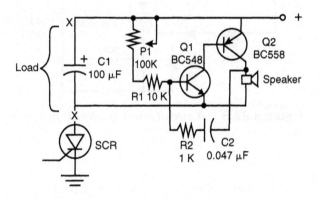

Figure 3.23.5 *An audio oscillator.*

Parts List—Audio Oscillator

- Q1 BC548 or equivalent general-purpose *negative-positive-negative* (NPN) silicon transistor
- Q2 BC558 or equivalent general-purpose *positive-negative-positive* (PNP) silicon transistor
- P1 100 kΩ trimmer potentiometer
- R1 10 kΩ × 1/8-watt resistor (brown, black, orange)

R2   1 kΩ × 1/8-watt resistor (brown, black, red)

C1   100 μF × 12-volt electrolytic capacitor

C2   0.047 μF ceramic or polyester capacitor

SPKR   4- or 8-ohm small speakers (2.5 to 10 cm)

Terminal strip, box, wires, solder, etc.

Driving an AC Lamp

The circuit shown in Figure 3.23.6 drives an incandescent lamp when plugged into the AC power supply line.

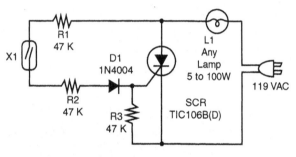

Figure 3.23.6   *A circuit driven by the AC power line.*

The builder must take care with this circuit because it is powered by the AC power line. The circuit must be inserted in a plastic or wooden box to avoid the possibility of causing shock. Remember that the sensor is also connected to the power supply line and therefore needs to be completely protected from accidental contact with the terminals. The circuit is very simple and can be mounted using a terminal strip, as shown by Figure 3.23.7.

This circuit is not self-latching, so when the magnet is removed the SCR will automatically turn off. The SCR must be installed on a small heatsink, and the lamp should be limited to 300 watts.

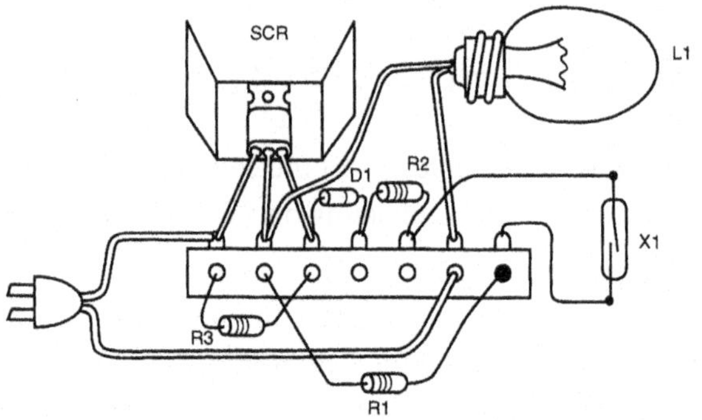

Figure 3.23.7   *The circuit mounted on a terminal strip.*

Parts List—Driving an AC Lamp

SCR   TIC106B, MCR106-4 or equivalent SCR (110 AC power line with suffix D or 6 for the 220/240 AC power line)

D1   1N4004 silicon rectifier diode

R1, R2, R3   47 kΩ × 1/8-watt resistors (yellow, violet, orange)

X1   Reed switch

Terminal strip, wires, heatsink for the SCR, wires, power cord, etc.

The Technology Today

Although many timers used in automatic kitchen devices are mechanical, a greater number of high-accuracy timers are electronic. The mechanical or electromechanical timers found in many appliances today have two operating principles.

The simple ones use a spiral string, such as those in antique clocks, to move a mechanism. When adjusting the scale to the desired time, the spring stores enough energy to move the mechanism. Another common timer found today is one that uses an electric motor to drive the mechanism. The motor has an accurate gear reduction system that drives a switch at a determined time period.

Ideas to Explore

Of course, the challenge proposed in this project is only one of many possible ideas for applications of the circuit. Some additional ideas are given as follows:

- Use the project to add automatic devices that use reed switches and magnets.
- Implement experiments in physics class using reed switches and magnets.
- Make an alarm by replacing the lamp with a buzzer and installing the reed switch (one or more) in a door or window.

Project 24—Experimental PLC

Programmable logic controllers, or PLCs, are primarily used in the control of industrial processes. A PLC is a small box attached to an industrial machine in which all the manufacturing operations of the machine are programmed. The advantage of the PLC is that the same machine can be used to manufacture different objects as different sets of instructions can be stored in each PLC. The operator simply changes the program of the PLC for the new task.

Robots and mechatronic devices, because they involve programmed operations, can be controlled by PLCs. Of course, inexpensive PLCs can be used in experimental robots or can learn the special language used to program the robots. However, we propose the construction of a very simple PLC to teach the evil genius how PLCs work and how they can be used to control small mechatronic automated devices.

The circuit does not need special computer language to be programmed because the program is made for specific hardware. The number of operations that can be performed is limited but is great enough to make interesting devices.

Objectives

- Learn how a PLC works.
- Use a PLC to control simple tasks.
- Create automatic routines controlled by a PLC.

How It Works

A real PLC is a small box containing a microcontroller as well as inputs and outputs, as shown in Figure 3.24.1. Wired to the inputs are sensors or a keyboard that sends information to the PLC about an external situation that should be considered while the program is running. The microcontroller receives this information, and while the program is running a PLC determines the next move to be made. For instance, it might be programmed to turn off a motor if a lever is low or turn on the motor if the lever is high.

Figure 3.24.1   *A common PLC.*

The decisions are dependent on many variables. The motor is turned on when the lever is low, but the motor can also be turned on by a key or from some other kind of actuator. The outputs send the appropriate signal to the device(s) to be commanded.

The program is written in a special language and stored in the PLC's memory. Many standard languages are adopted for use by the different types of PLCs. The PLC produces signals to respond to the sensors that are activated according to a program stored inside its memory.

Section Three   The Projects

The most important characteristic of a PLC is that it can be used to control any process. As long as we know what we want from the machine, we can program that into the PLC. PLCs can be used to control everything from small automations in a home to complex industrial machines.

Our project is not a real PLC, although it operates following the same principles. To make it easy to build using inexpensive parts, the circuit is very simple, as shown by Figure 3.24.2.

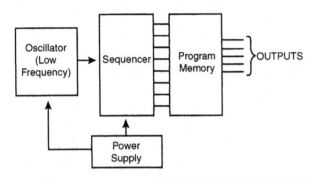

Figure 3.24.2 *Block diagram for our PLC.*

The input block controls a simple oscillator that can be programmed to determine the speed of the process. This block uses a 555 *integrated circuit* (IC), and C1 determines the range of speeds for the PLC. Values between 10 and 1,000 µF are recommended. Using a 1,000 µF capacitor, the execution time of each line of the program when P1 is in the maximum resistance is about 15 minutes each.

The oscillator is controlled by an external sensor and has two additional inputs for commands: an *enable* (EN) input that stops the oscillator and a *reset* (RST) input that returns the count of the Johnson counter in the second block to zero.

This block provides pulses to the second block, a Johnson counter with 10 outputs, activated in the 1 of 10 mode. A 4017 IC is used and it determines the size of the program memory: 10 lines or 10 words. The number of bits depends on the following block.

The next block is the program memory and is made with diodes. This block is formed by horizontal lines that are wired to the outputs of the sequencer block and vertical lines that correspond to the outputs.

When we want a high logic level in the second line to cause the first and second outputs to go high, simply connect the diodes, as shown in Figure 3.24.3.

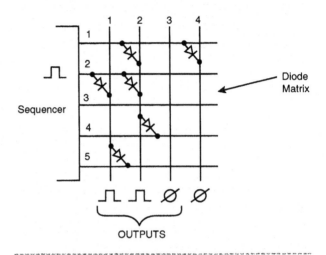

Figure 3.24.3 *Programming the matrix.*

So the number of diodes and their positions determine when each output is activated by one of the program lines. The diodes are mounted in a small *printed circuit board* (PCB) and plugged into a connector. The board *is* the program. The outputs can activate relays or other circuits as suggested by Figure 3.24.4.

The circuit can be powered by a 5- to 12-volt source according to the loads to be controlled by the outputs. But remember that the outputs of the matrix can provide only 0.88 mA when powered by a 5-volt supply, 2.25 mA with a 10-volt power supply, and 8.8 mA with 15-volt supply. These currents determine the number of output lines for the PLC.

How to Build

Figure 3.24.5 shows the schematic for the PLC, which can be mounted on a PCB as shown in Figure 3.24.6. In the same figure, we show the details of the program board where the diodes are soldered according to the program. The reader must have more than one of these boards to make different programs for the PLC.

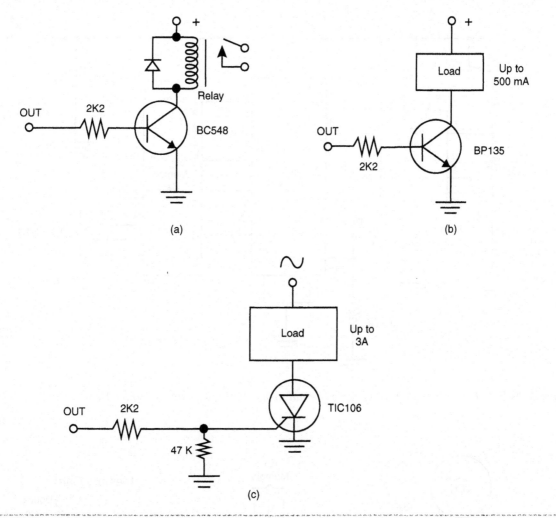

Figure 3.24.4 *Activating relays from the outputs.*

The circuit can be powered by AA cells or a power supply with voltages ranging from 5 to 12 volts. Figure 3.24.7 shows a PCB for relays.

When mounting, take care with the polarized components such as the ICs, diodes, electrolytic capacitor, and the power supply.

Testing and Using

To test your PLC, you can power an experimental circuit formed by *light-emitting diodes* (LEDs) plugged into the outputs, as shown in Figure 3.24.8.

After inserting the program card (the board with the diodes) and powering the circuit, adjust P1 to the minimum resistance. The LEDs will glow according to the program determined for the diodes. Once you verify the operation, you can use your PLCs to control automated devices or for demonstrations.

Parts List—Experimental PLC

IC-1	555 IC timer
IC	4017 *complementary metal oxide semiconductor* (CMOS) IC
D1 to Dn	1N4148 general-purpose silicon diodes (see text)
P1	1 MΩ trimmer potentiometer
R1, R2	2.2 kΩ × 1/8-watt resistors (red, red, red)
R3	100 kΩ × 1/8-watt resistor (brown, black, yellow)

Section Three   The Projects

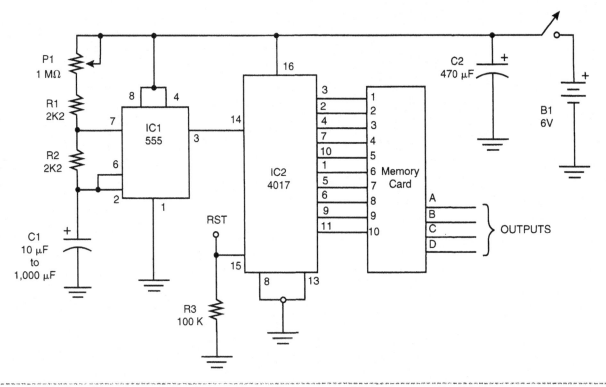

Figure 3.24.5 *The schematic for the experimental programmable logic controller PLC.*

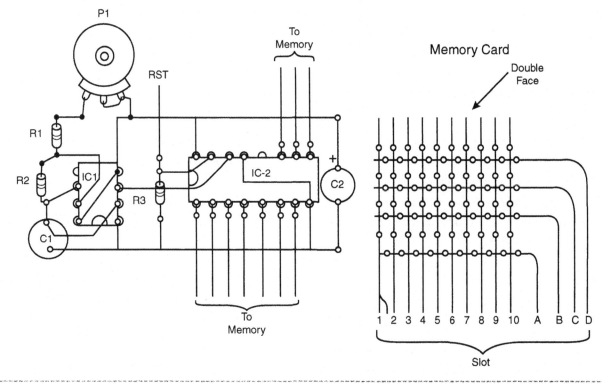

Figure 3.24.6 *The circuit can be mounted on a PCB.*

198   Mechatronics for the Evil Genius

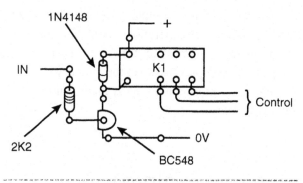

Figure 3.24.7   *A PCB for relays.*

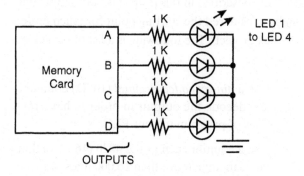

Figure 3.24.8   *A test circuit using LEDs.*

C1   10 μF to 1,000 μF × 12-volt electrolytic capacitor

C2   470 μF × 16-volt electrolytic capacitor

B1   4 AA cells or other power supply totaling 6 volts (see text)

Power supply, PCB, slide connector (according to the lines of program and outputs), wires, solder, etc.

Cross Themes

Students in industrial automation and engineering courses study PLCs, but, of course, they study these devices in much greater detail. In high-school-level courses, it is possible to include lessons on this type of device in elective subjects that study mechatronics. The idea is to study a simple automated process that could be easy to understand even for persons who don't have the necessary knowledge in electronics. Therefore, you can use this PLC in cross themes such as the ones suggested in the following list:

- Study how an automated process functions.
- Make automated devices for demonstrations.
- Teach applications of digital logic.

Additional Circuits and Ideas

Adding a Sensor-Activated Switch

Figure 3.24.9 shows how to modify the oscillator to operate as a monostable device triggering the counter from the signals of a sensor.

Each time the sensor is closed the monostable device produces a pulse and the counter advances one line in the memory. Thus, the sequence of operations programmed in the memory is controlled by the sensor. The sensor can be a reed switch, a pushbutton, or another kind of on/off sensor.

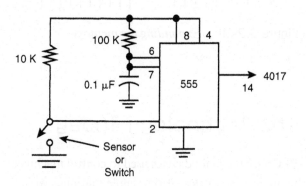

Figure 3.24.9   *Adding a sensor.*

Adding More Logic

An interesting way to add more logic to the PLC is shown in Figure 3.24.10. Four gates of a 4093 are added to the circuit to make the outputs dependent on the logic levels of inputs A to D. These inputs can be connected to sensors, logic arrays, or other devices to command the PLC.

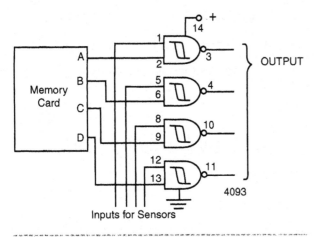

Figure 3.24.10 *Adding more logic.*

Expanding the Memory

Finally, you can have more than 10 lines of program memory in your PLC using two or more 4017s, as shown in Figure 3.24.11.

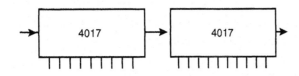

Figure 3.24.11 *Expanding the memory.*

The Technology Today

PLCs are the main devices used to control machines in modern industries—from simple machines with only a few functions to complex machines that make hundreds of different operations. Some of these devices have the memory to store large programs that perform complex tasks, yet the memory is dependent on information sent by many sensors or other data sources.

Ideas to Explore

The PLC described in this project is very simple but can be considered the starting point for much more complex projects. The following is a list of some ideas to explore:

- Use a *binary coded decimal* (BCD) counter and decode the outputs in order to have 16 lines of program in your PLC.

- Use a counter such as the 4020 (16 k) so that you can have more than 16,000 lines of program. Of course, you'll need a decodification block.

- Try to use RAM instead of a diode matrix and program it using a computer.

- Develop an automated device controlled by your PLC. Try to add it to the autonomous robot (refer to Project 17), giving it some intelligence.

Project 25 — The Mechatronic Talking Head

Imagine a mechatronic head wired to a circuit that makes it talk. Of course, the head doesn't actually talk. A trick involving electronics and the skills of the builder will be responsible for this. The idea is that a circuit moves the mouth of the head while it is receiving sound from a radio receiver, and in this case the sound will be something that you or a friend will say through a wireless microphone. You or your friend will also be hiding out of sight, so the people watching the head will think that it is talking.

The aim of this project is to build the head and the circuit that will move the mouth on cue from the sound picked up by the receiver. In the basic version, the head is very simple, but the evil genius who may have skills with building materials such as plastic, rubber, and other materials can build a head with a clear resemblance to a human head. The circuit uses only a common FM receiver and a wireless microphone, and the parts that move the head use very few components. They can be built even by the reader who has little experience with electronics.

You are encouraged to explore different ideas of using your talking head. Plug the head into a computer and make it speak from a program recorded in the computer. Change the voice so it sounds like a computer or robot voice to add a little robotic realism to your project.

Objectives

- Build an experimental head that talks.
- Work with a circuit that converts sounds into movement or light.
- Create a talking toy.
- Design a new and terrifying ghost for Halloween.
- Study movements and sounds.

How It Works

Figure 3.25.1 shows the block diagram for this project. The sounds picked up by the wireless microphone are transmitted to a small portable FM receiver that fits inside the head. Sound signals consist of frequency and amplitude changes in a voltage. By adjusting the level of the trigger of a *silicon controlled rectifier* (SCR), we can make it turn on at the peaks of the signals, as shown in Figure 3.25.2.

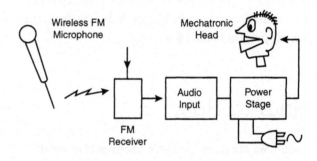

Figure 3.25.1 *Block diagram for the mechatronic talking head.*

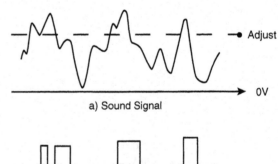

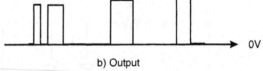

Figure 3.25.2 *Triggering an SCR at the peaks of the signal.*

This means that the changes in the voice level of a person who talks into the microphone will make the SCR turn on and off. A lamp plugged into the SCR will blink according to the person's voice. This is just

Project 25 – The Mechatronic Talking Head

one of the applications for the circuit—converting sound into light.

But what we hope to accomplish with our project is more than that. Instead of using a lamp, we are going to control a solenoid attached to a mechanism that moves the head's mouth. When we talk into the microphone, the changes in the voice level will push and pull the mechanism, moving the head's mouth at the same rhythm as the words being spoken into the mic. How realistic the talking head appears will depend on many factors, such as the adjustments, the mechanism attached to the mouth, and how the reader will use your talking head.

The circuit to convert the signals picked up by the FM receiver is very simple because we use an output made from a phone jack. SCRs such as the 106 are very sensitive and can be triggered directly from the signals present in the phone jack.

How to Build

Again, we are going to divide the project into two parts: (1) the electronic part and (2) the mechanical part.

Electronics

Figure 3.25.3 shows the schematic for the electronics of the mechatronic talking head.

Because the circuit is very simple and uses few parts, they can be soldered to a terminal strip, as shown in Figure 3.25.4.

Notice that the circuit is powered directly from the AC power line. No isolation transformer is used, so you must take care to protect all the parts against accidental touches, avoiding shock hazards.

A wooden box can be used to house the circuit and the head can be mounted on it. When mounting, be careful with the polarized parts, such as the SCR and diodes. Because the SCR will not power large loads, it does not need to be mounted on a heatsink. The lamps are low-power, 5-watt lamps, and the transformer with the solenoid doesn't require more than 20 watts. However, if you intend to power heavier loads such as more powerful lamps, attach the SCR to a heatsink.

The input transformer (T1) is any small, low-voltage transformer with a secondary coil rated for 5 to 12 volts and for currents in the range of 150 to 500 mA. The primary coil is 117 VAC. This transformer is not used in the original function, which is to lower the voltage of the AC power line, but as an

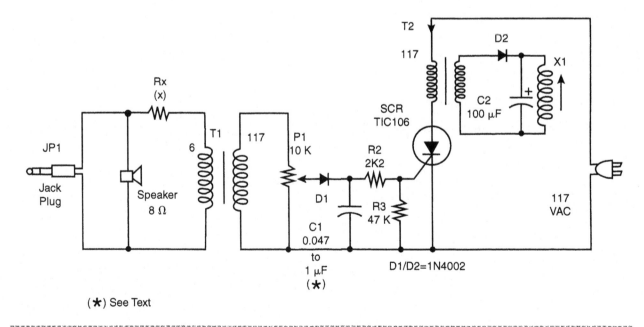

Figure 3.25.3 *The schematic for the mechatronic talking head.*

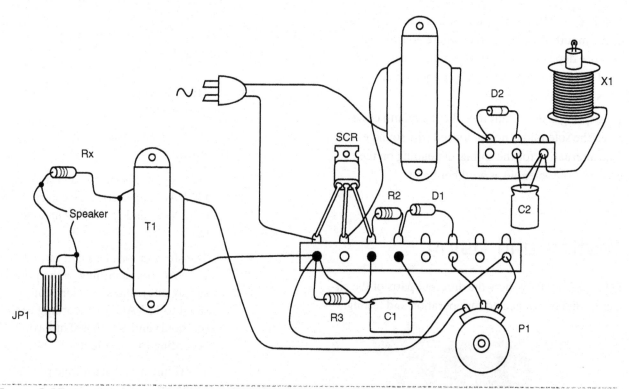

Figure 3.25.4 *Mounting using a terminal strip as a chassis.*

isolation transformer and to match the audio signals from the output of the FM receiver.

C1 determines the response of the circuit to bass and treble. Adjusting to values between 0.047 and 1 µF can provide the best performance.

T2 is a transformer with a 110 VAC primary coil and a 12-volt × 800 mA secondary coil, according to the solenoid being used.

The solenoid X1 is any small 6- to 24-volt solenoid with a current range from 150 to 500 mA. Solenoids or motors with 117 VAC are also acceptable.

PL1 is a plug chosen according to the output of the FM radio used as the receiver. Typically, this would be a mono plug, but if the receiver has a stereo output, the adaptation shown in Figure 3.25.5 must be made.

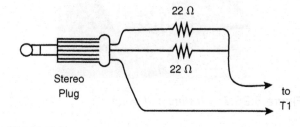

Figure 3.25.5 *Using a stereo receiver.*

ing the mechanism to move his or her mouth. Figure 3.25.6 shows a very simple solution for the mechatronic talking head using cardboard, wood, and other simple materials.

Mechanics

The basic mechanics are very simple. Of course, the reader who has existing mechanical skills might choose to upgrade this version. For example, you might consider using a plastic or rubber mask with the likeness of some famous person and then attach-

Figure 3.25.6 *Basic construction for the mechatronic talking head.*

The core of the solenoid moves the lower portion of the mouth. The exact point where you will couple the core depends on the power of the solenoid. You'll need to experiment to find the coupling point that gives the best performance.

A spring is important to make the mouth close when the solenoid is not energized (the sound ends). One alternative for the spring, depending on the power of the solenoid, is a rubber band.

Testing and Using

Figure 3.25.7 shows how the different parts of the circuit should be connected for operation and testing.

Figure 3.25.7 *Testing and using.*

First, tune the FM receiver to find a free point on the dial. Adjust the FM wireless microphone to pick up the signal. Place the receiver on medium volume so that the listener or viewer can hear your voice (the voice of the talking head) loudly and clearly. Now, while speaking into the microphone, adjust P1 on the head circuit to cause the solenoid to move the mouth. The circuit is ready for use.

Make adjustments to the solenoid and mechanical portion if necessary or change C1 if you want the head speaking with low-frequency or high-frequency tones.

Parts List—Mechatronic Talking Head

SCR	TIC106B (117 VAC) or TIC106B (220/240 VAC) SCR
D1, D2	1N4002 or equivalent silicon rectifier diodes
T1	Transformer: primary coil rated to 117 VAC, and secondary coil rated 5 to 12 VAC and 150 to 500 mA (see text)
T2	Transformer: primary coil rated to 117 VAC (or according to the power supply line), and secondary coil rated to 12 VAC (or according to the solenoid) and 500 to 800 mA (or according to the solenoid)
P1	10 kΩ linear or logarithmic potentiometer
R1	2.2 kΩ × 1/8-watt resistor (red, red, red)
R2	47 kΩ × 1/8-watt resistor (yellow, violet, orange)
C1	0.047 µF to 1 ceramic or polyester capacitor (see text)
C2	100 µF × 16-volt electrolytic capacitor
L1	117 VAC × 5-watt miniature incandescent lamp
X1	Small 12- to 24-pull solenoid (see text)
SPKR	8-ohm × 10 cm loudspeaker

Power cord, terminal strip, wires, solder, etc.

Cross Themes

Any elementary or high school project that includes movement and speech probably will cross over with this project. Some ideas are given in the following list:

- Use the project to teach spelling to children.

- Study movement.
- See how sounds can be converted into movement and light.

New Circuits and Ideas

This project can be altered in many ways. You don't need to build it exactly as described. A few suggestions for alternatives are given here.

Using an Audio Amplifier

Instead of sending the voice to the head via a radio signal, use wires and a low-power amplifier, as shown in Figure 3.25.8. Any amplifier with a power of 100 mW or greater can be used to activate the head. The necessary connections are shown in the figure.

Take care to add the power-limiting resistor according to the output power of your amplifier. The following table provides the approximated value for this resistor. The values can change because many amplifiers are not rated by the *real power* (RMS) output but by the *peak power* (PMPO).

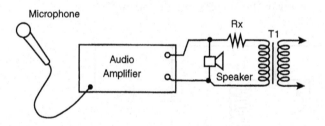

Figure 3.25.8   *Using an audio amplifier.*

Output Power	Rx
0 to 1 W	None
1 to 5 W	10 V × 1 W
5 to 20 W	22 V × 1 W
20 to 50 W	47 V × 2 W
50 to 100 W	100 V × 2 W

Using a Rotary Solenoid or a Small DC Motor with Gears

Rotary solenoids and small DC motors with gearboxes can be used to move the mouth of the head, as shown in Figure 3.25.9.

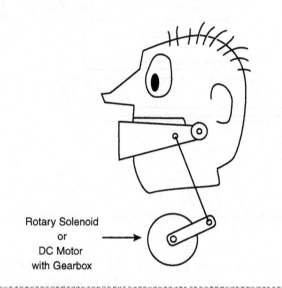

Figure 3.25.9   *Using rotary solenoids or DC motors with gearboxes.*

The figure shows how the shaft of the motor can be used to produce linear movements that open and close the mouth. Note that, for the best performance, the movement should be made at low speeds.

Using a 117 VAC Motor

If you have a 117 VAC motor with a gearbox and a shaft that moves fairly slowly, you can use it. The connection is easier because the transformer T2 can be omitted.

Using the Computer

The output from a computer's loudspeakers can be used to drive this circuit, as shown in Figure 3.25.10. You can use the output coupled with the input of an audio amplifier (suggested earlier) and the amplifier to drive the head.

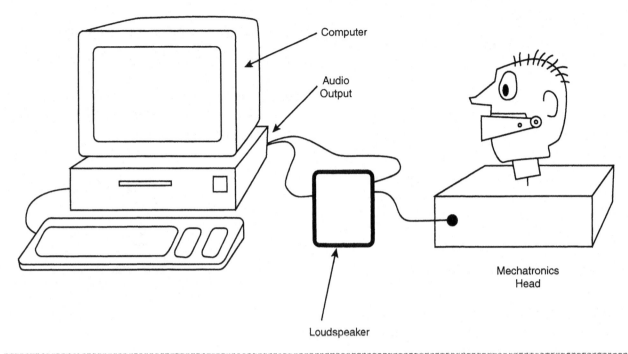

Figure 3.25.10 *Using the audio output of a computer to drive the circuit.*

You can record on the computer what the head will say and even add sound effects to make a "computer voice," for instance. The phrases can be activated by a touch of the keyboard or a click of the mouse.

A Small FM Transmitter

If you don't have a small FM transmitter to be used as a wireless microphone, you can build one. Figure 3.25.11 shows the schematic for a one-transistor transmitter that can send the signals to a small FM receiver at distances of up to 50 meters.

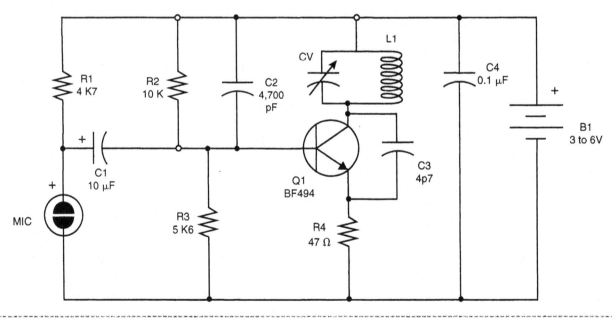

Figure 3.25.11 *A small FM transmitter.*

Figure 3.25.12 shows the *printed circuit board* (PCB) for this project.

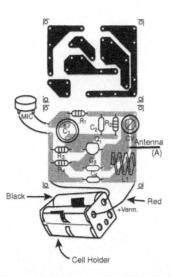

Figure 3.25.12 *PCB for the transmitter.*

L1 is formed by four turns of any enameled wire of 20 to 28 gauge, with a diameter of 1 centimeter, and without a core.

The antenna is a piece of bare wire 15 to 30 centimeters long. The circuit with cells is housed in a small plastic box.

Using the FM transmitter is very simple. Adjust CV to find a free point on the FM band and then speak into the microphone. If the signals disappear when you are only a few meters away from the receiver, find another frequency. You may have found a secondary or spurious signal, not the primary signal. If the receiver produces a strong whistle when you are talking, place the microphone farther away from the receiver to avoid the acoustic feedback.

Parts List—Using a Small FM Transmitter

- Q1 BF494 or the equivalent RF silicon *negative-positive-negative* (NPN) transistor
- MIC Two-terminal electric microphone
- R1 4.7 kΩ × 1/8-watt resistor (yellow, violet, red)
- R2 10 kΩ × 1/8-watt resistor (brown, black, orange)
- R3 5.6 kΩ × 1/8-watt resistor (blue, green, red)
- R4 47 Ω × 1/8-watt resistor (yellow, violet, black)
- C1 10 μF × 12-volt electrolytic capacitor
- C2 4,700 *pico farads* (pF) ceramic capacitor
- C3 4.7 pF ceramic capacitor
- C4 0.1 μF ceramic capacitor
- CV Trimmer capacitor (15 to 40 pF of maximum capacitance)
- B1 2 or 4 AA cells and holder (totaling 3 or 6 volts)
- L1 Coil (see text)
- A Antenna

PCB, plastic box, wires, solder, etc.

The Technology Today

Many labs around the world in research centers and universities are working on the "talking head" not only for use in robots or other machines, but also to study diseases related to human speech. In the future, we may find these talking heads in many places actually talking with us aided by some voice-recognition system so they understand what we are saying.

Ideas to Explore

A talking head or mechatronic head can be used in many interesting projects, as previously suggested. You might also explore the following:

- Add a transmitter to the head to allow the person who is speaking into the microphone to also hear what is being said to the head by the listeners, even when the transmitting person is far away.

- Create a terrific talking ghost for your next Halloween using the same principles found in this project.
- Create a robot controlled by a computer that has a talking head, using the principles described here.

Index

117 VAC motors, 205
555 integrated circuits (ICs)
    and analog computers, 114–116
    in elevators, 133, 134
    and light-beam remote control, 180
    and Lissajous figures, 96
    in magic motion machines, 146–147
    and programmable logic controllers, 196
    and pulse width modulation, 35
    in touch-controlled motors, 119–120, 125–126
4013 CMOS D-type flip-flop, 134–135
4093 CMOS integrated circuits (ICs), 125–126, 149
4093 integrated circuits (ICs), 102–103, 199

A

Airboats, 183–187
Alternative energy sources, 73–74
Amplifiers
    and analog computers, 106–107, 116–117
    for galvanometers, 57–59
    and Lissajous figures, 104–105
    and talking heads, 205–206
Analog multimeters, 175
Analog power control, 67; *see also* Potentiometers
Anemometers, 79–80
Audio amplifiers, 205–206
Audio generators, 101
Audio oscillators, 193–194

B

Ballistics, 89
Battery chargers, 77–78
BD135 motors, 126, 139
Beaufort scale, 80
Bottle's challenge, 63–64

C

Cadmium sulfide (CdS), 178
Cameras, 143
Cannons, electronic, 83–93
    battle, 90
    building, testing, and using, 86–88
    power, 85–86
    variations and experiments, 88–93
Capacitative discharge circuit, 189–190
Capacitors
    in coin tossers, 189
    in diagrams, 9
    and electronic cannons, 83–86, 89–90, 92
    in eolic generators, 76
    in ionic motors, 42, 49, 52
    and pulse width modulation, 35, 36
    in robots with sensors, 159
    in stepper motors, 141
Catapults, 91–92, 190
Coils
    in coin tossers, 188, 189
    and electromagnets, 62
    and electronic cannons, 84–85, 87–88, 91
    in galvanometers, 53
    in ionic motors, 42
    in magic motion machines, 148
    in nerve test, 152, 153
    in stepper motors, 137
Coin tossers, 187–190
Compact discs (CDs), 12, 15, 22, 24–26
Comparators, 117–118, 175–177
Compasses, 58–59
Complementary metal oxide semiconductor (CMOS) flip-flop integrated circuits (ICs) 4013, 181
Complementary metal oxide semiconductor (CMOS) integrated circuits (ICs) 4017, 144, 200
Complementary metal oxide semiconductor (CMOS) integrated circuits (ICs) 4093, 48
Component testers, 58
Computers; *see also* Computers, analog
    and elevators, 132
    and Lissajous figures, 101
    and pulse width modulation, 39
    and talking heads, 205–206
Computers, analog, 106–118
    project, 108–113
    purpose, 107–108
    scales, 108–113
    testing and using, 113
    variations and experiments, 113–118
Constant current source, 72–73, 166–167, 169–170
Cranes, 63
Cross themes, 3
    and airboats, 185
    and analog computers, 114
    and coin tossers, 189
    and electromagnets, 64
    and elevators, 132–133
    and eolic generators, 81
    and galvanometers, 53, 56–57
    and ionic motors, 48
    and light-beam remote control, 179–180
    and Lissajous figures, 101–102
    and magic motion machines, 149
    and nerve test, 156

and position sensors, 175
and potentiometers, 70
and programmable logic controllers, 199
and race cars, 19
and RobCom, 28
and robots with sensors, 162
and SMA robotic arms, 169
and stepper motors, 143
and talking heads, 204–205
and timers, 193
and touch-controlled motors, 122–123
Cybernetics, 1

D

Darlington transistors
    and potentiometers, 71–72
    and pulse width modulation, 37–38
    in race cars, 13, 14
    in RobCom, 30–32
    and stepper motors, 143–144
DC motors; *see also* Touch-controlled motors
    and airboats, 183
    diagram, 9–10
    in elevators, 127–129, 131
    and eolic generators, 74–77
    and Lissajous figures, 100
    in nerve test, 153
    and potentiometers, 72–73
    and pulse width modulation, 33, 35–37, 39
    in race cars, 11
    in RobCom, 22, 24–26, 29–30
    and talking heads, 205
Death circuit, 30
Diagrams, 9–10
Digital multimeters, 175
Digital-to-analog converters (DACs), 39
Diodes; *see also* Light-emitting diodes (LEDs); Silicon diodes for alternating current (SIDACs)
    in analog computers, 111
    and electronic cannons, 84
    in eolic generators, 76
    and programmable logic controllers, 196
    in robots with sensors, 159
Double pull/double-throw (DPDT) switch, 30
Dual-in-line (DIL) relays, 159
Dynamos, 75–76, 81, 82

E

Electrofluid dynamics (EHD), 41
Electromagnets, 61–66; *see also* Magnetic fields
    building, testing, and using, 62–63
    and Lissajous figures, 101
    variations and experiments, 63–66
Electronics, educational, 2
Electrostatic experiments, 47–48
Elevators, 127–136
    building, testing, and using, 129–131
    competition, 133
    electronic circuit, 128–129
    mechanical part, 129

and potentiometers, 70
and pulse width modulation, 37
and touch-controlled motors, 126–127
variations and experiments, 131–136
Engineering, 3
Eolic generators, 73–82
    building, testing, and using, 76–77
    and galvanometers, 57
    variations and experiments, 81–82

F

Fans
    and airboats, 183
    and eolic generators, 74, 76, 82
    and pulse width modulation, 37
    in race cars, 12, 15, 18
    and touch-controlled motors, 122
Flip-flops, 134–135, 181
Flyback transformers, 42–44, 48, 49, 51
FM transmitters, 206–207
Function generators, 101

G

Galvanometers, 53–61; *see also* Magic motion machines
    building and testing, 55–56
    definition, 54
    and eolic generators, 80
    increasing sensitivity, 57–58
    variations and experiments, 56–61
Gearboxes
    in analog computers, 117
    in elevators, 127, 131, 132
    in RobCom, 32
    in robots with sensors, 159, 162
    in talking heads, 205
Gears, 12, 16–19, 143
Ground connection (GND), 124

H

Hand-drying machines, 122
H-bridge, 30–32, 134
Heaters, 70
Heatsinks, 14, 36, 42
High-power amplifiers, 105
High-power circuits, 48–50
High-voltage circuits, 40, 42, 44–52; *see also* Nerve test
High-voltage inverters, 81–82
Hydrotronics, 191
Hysteresis, 165–166

I

Indicators
    in analog computers, 108–112, 116–117
    and position sensors, 172–175
    sound, 114–116
Input/output (IO), 39
Integrated circuits (ICs); *see also* specific circuits
    for elevators, 129

in robots with sensors, 159
and SMA robotic arms, 166
and soldering, 9
in stepper motors, 144
Inverters, 40, 42, 44, 156–157
Ionic motors, 40–52
building and testing circuit, 42–45
variations and experiments, 46–52

J

Jacob's Ladder, 47
Joule's law, 70
Joysticks, 28–29, 138–139

K

Kapec, Karel, 1

L

Lamps; *see also* Light
and eolic generators, 77, 81–82
and nerve test, 153–154
and potentiometers, 69
in spacecraft, 45–46
and timers, 194
wireless fluorescent, 46
Lasers, 143; *see also* Lissajous figures
Levers, 191
Lie detectors, 57, 58
Light; *see also* Lamps
automatic, 79
and ionic motor, 41
and potentiometers, 69
and pulse width modulation, 38–39
Light-beam remote control, 177–182
and airboats, 186
building, testing, and using, 178–179
variations and experiments, 180–182
Light-dependent resistors (LDRs)
and electronic cannons, 85, 90
and galvanometers, 59–60
and light-beam remote control, 178, 179
and pulse width modulation, 38–39
in race cars, 13, 14, 17
in robots with sensors, 162
in stepper motors, 139
Light-emitting diodes (LEDs)
and analog computers, 116–117
in coin tossers, 189
and electronic cannons, 86, 88, 89
and eolic generators, 77
and light-beam remote control, 179
in magic motion machines, 146, 148
and programmable logic controllers, 197
Light sensors, 162
Linear power control, 33–34, 64, 73, 133, 184–185
Linear scales, 112
Lissajous figures, 93–106
building circuit, 96–98
definition, 93–94

mechanical part, 98–99
project, 95–96
testing and using, 99
variations and experiments, 99–106
Lissajous, Jules Antoine, 94
LM350T integrated circuits (ICs), 72–73, 166, 167
Logarithmic scales, 112

M

Magic motion machines, 146–151; *see also* Galvanometers
building, testing, and using, 147–148
variations and experiments, 149–151
Magnetic fields, 61–62; *see also* Electromagnets; Magnets
and coin tossers, 188
and electronic cannons, 83, 84
and eolic generators, 74–75
and galvanometers, 53–54
Magnetic induction, 64, 146
Magnets, 128, 129, 147, 148, 156
MC1411 integrated circuits (ICs), 144
the mechanical advantage (TMA), 18
Mechatronics
definition and history, 1–2
tools and principles, 2–4
Metal-oxide semiconductor field-effect transistors (MOS-FETs), 19–20, 37
Microscopes, 69
Mine field construction, 127
Mirrors, rotating, 100
Mixers, 123
Monostable blocks, 158
Motion machines, *see* Magic motion machines; Perpetual-motion machines, mobile
Motor controls, 67; *see also* Pulse width modulation (PWM) control
Mounting, 5–10
Multimeters, 56, 70, 77, 175

N

Negative-positive-negative (NPN) transistors
in nerve test, 157
and potentiometers, 67, 68, 71
in robots with sensors, 159
in stepper motors, 138–139
Negative temperature coefficient (NTC), 139
Nerve test, 152–158
building, testing, and using, 154–156
premounting experiments, 153–154
variations and experiments, 156–158
Newton's color wheel, 123, 143
Newton's Laws, 18, 41
Nickel cadmium (NiCAD) cells, 77–78
Nitinol, 164–165

O

Oesterd, Hans Christian, 53–54, 61
Ohm's law, 89, 166
Operational amplifiers, 116–117
Operator in analog computer, 108

Oscillators
   in analog computers, 114–115
   in ionic motors, 42, 48, 50–52
   and Lissajous figures, 102–104
   in magic motion machines, 150–151
   in nerve test, 157
   and programmable logic controllers, 196
   in stepper motors, 138, 139, 141
   and timers, 193–194

P

Pendulums, 191
Perpetual-motion machines, mobile, 60–61
Photocells, 57
Piezoelectric transducers, 114–116
Plotters, 142
Point effect, 41
Position sensors, 172–177
   building, testing, and using, 173–174
   variations and experiments, 175–177
Positive-negative-positive (PNP) transistors, 35, 42, 71, 147
Potentiometers, 67–73
   and analog computers, 106, 108–112, 117, 118
   building and using, 68–69
   and electromagnets, 64
   and electronic cannons, 90
   in elevators, 129, 133
   in galvanometers, 55–56, 58
   in ionic motors, 42, 45–46
   and Lissajous figures, 97, 100
   and position sensors, 172–174
   and pulse width modulation, 36, 38
   in robots with sensors, 162
   and SMA robotic arms, 166, 170
   in touch-controlled motors, 120
   variations and experiments, 69–73
Printed circuit boards (PCBs), 6–8
   and analog computers, 114, 115
   in elevators, 129
   and experimental radios, 78
   and ionic motors, 50
   and Lissajous figures, 96
   and programmable logic controllers, 196
   and pulse width modulation, 35
   in robots with sensors, 159
   in stepper motors, 139
   and talking heads, 207
   and terminal strips, 6
   in touch-controlled motors, 120
Programmable logic controllers (PLCs), 195–200
   building, testing, and using, 196–199
   variations and experiments, 199–200
Projects, choosing, 4
Propellers, 12, 187
Pulleys, differential, 129, 132–133
Pulsed sources, 170–171
Pulse generators, 149–151
Pulses, 137–139, 140, 146, 147
Pulse width control (PWC), 128
Pulse width modulation (PWM) control, 33–39
   in airboats, 184–185
   building, testing, and using, 35–37
   definition, 33, 34
   and electromagnets, 63–65
   in elevators, 128, 129
   and Lissajous figures, 100
   operation principle, 34
   and potentiometers, 67, 73
   in RobCom, 29
   in robots with sensors, 162
   and SMA robotic arms, 166, 170–171
   variations and experiments, 37–39

R

Race cars, 11–21
   building and testing, 13–17
   competition, 17–18
   and stepper motors, 136
   and touch-controlled motors, 122, 123
   variations and experiments, 18–21
Radios, experimental, 78–79
Reduction systems, 127, 129, 131
Reed switches
   in elevators, 128, 129, 134
   in nerve test, 156
   in stepper motors, 140
   and timers, 191–192
Relaxation circuits, 50–51
Relays, 156–157, 159, 175–176, 178–179, 196–197
Remote control, 22–24, 32, 89–91, 185–186; *see also* Light-beam remote control
Resistance, 173, Skin resistance
Resistors, 9–10, 49, 64–67, 125; *see also* Light-dependent resistors (LDRs)
Rheostats, 33–34, 67
RobCom, 22–32
   building and testing, 24–27
   competition, 23, 27
   definition, 22–23
   variations and experiments, 28–32
   weapons, 29–30, 83
Robotic arms, 142; *see also* Robotic arms, SMA experimental
Robotic arms, SMA experimental, 164–172
   building, testing, and using, 167–169
   project, 167
   variations and experiments, 169–171
Robotics; *see also* RobCom; Robotic arms, SMA experimental; Robots with sensors
   applications, 2
   definition and history, 1–2
   and programmable logic controllers, 195
   and pulse width modulation, 38–39
   and stepper motors, 136
   and touch-controlled motors, 122
Robots with sensors, 158–164
   building, testing, and using, 159–161
   variations and experiments, 162–164
R.U.R., Rassum's Universal Robots (Kapec), 1

S

Saturation, 13
Schematic diagrams, 9–10
Scientific method, 3–4
Sensors; *see also* Position sensors; Robots with sensors; Touch-controlled motors
    in elevators, 128, 129, 132, 134
    and light-beam remote control, 178
    in programmable logic controllers, 199
    in stepper motors, 139
Sequencers, 140–142, 144–145
Shape memory alloy (SMA), 142, 164–165, 171–172; *see also* Robotic arms, SMA experimental
Shock, *see* Nerve test
Silicon controlled rectifiers (SCRs)
    and coin tossers, 189
    and electronic cannons, 85–87
    and ionic motors, 50–51
    and talking heads, 201–202
    and timers, 192–194
    and touch-controlled motors, 123–124
Silicon diodes for alternating current (SIDACs), 52
Simple regulated power supply, 121
Sine scales, 112
Sine wave oscillators, 103–104
Skin effect, 41
Skin resistance, 58, 118–120
Slide rules, 107
Soldering, 5–8, 152
Soldering guns, 8–9
Soldering irons, 7–8
Solderless boards, 96, 178
Solenoids
    and airboats, 186
    and coin tossers, 187–190
    and electronic cannons, 84, 91–93
    and Lissajous figures, 100
    and stepper motors, 142
    and talking heads, 202–205
Sound effects, 30, 163, 186–187
Sound indicators, 114–116
Spacecraft, 41–43, 45–46, 52
Standing waves, 143
Stepper motors, 136–145
    building, testing, and using, 139–142
    in elevators, 132
    and Lissajous figures, 100–101
    operating principle, 136–138
    project, 138
    sequence generator for, 140–142
    variations and experiments, 142–145
Summing circuits, 114
Surface mount devices (SMDs), 5, 7
Surface mount technology, 5
Switched mode power supply (SMPS), *see* Pulse width modulation (PWM) control
Switches, 138–139, 141, 199; *see also* reed switches

T

Talking heads, 201–208
    building, testing, and using, 202–204
    and stepper motors, 142
    variations and experiments, 205–208
Technology
    applications, 2–3
    in projects, 5–10
Terminal strips, 5–6
    and electromagnets, 65
    for electronic cannons, 86
    for experimental radios, 78
    for galvanometers, 57
    and potentiometers, 68
    in race cars, 14
    for talking heads, 202
    and timers, 192
Thunderbolts, domestic, 47
Time constant, 89
Timed circuits, 20–21
Timers, 191–195
    in elevators, 133–134
    and light-beam remote control, 180–181
    in robots with sensors, 162–163
Tools, 9
Touch-controlled motors, 118–127
    building, testing, and using, 120–121
    variations and experiments, 121–127
Transducers, *see* Potentiometers
Transformers, 82, 86–87, 152–153; *see also* Flyback transformers
Transistor audio amplifiers, 104
Transistors; *see also* specific transistors
    in diagrams, 10
    and galvanometers, 57, 59
    in ionic motors, 42, 49
    and pulse width modulation, 33, 34, 36, 37
    in robots with sensors, 159
    and SMA robotic arms, 169–170
Traps, 122–123

U

ULN2001A integrated circuits (ICs), 144

V

Very high-voltage (VHV), 40
Vibrating blades, 101
Voltage in diagrams, 10
Voltage regulators, 81

W

Waveform, 10
Wind, *see* Eolic generators
Window comparator, 176
Windows, automatic, 37, 121–122

PICAXE EXPERIMENTER STARTER KIT

READER OFFER!

Revolution Education are offering a special price PICAXE experimenter starter kit to every reader of 'Programming and Customising the PICAXE Microcontroller'.

The experimenter kit includes everything you need to get started with the PICAXE system - a PICAXE-08M chip, experimenter PCB, serial download cable, battery box and all the components required to populate the PCB. Once assembled the kit provides a versatile demonstration board for carrying out many of the experiments from this book.

For more advanced experiments the board can also be linked to an external prototyping breadboard (not included).

To purchase the kit visit the PICAXE website at

www.picaxe.co.uk

and type this key phrase into the search window

Prog Custom Kit